Clean Synthesis Using Porous Inorganic Solid Catalysts and Supported Reagents

RSC Clean Technology Monographs

Series Editor: J.H. Clark, *University of York, UK*

Advisory Panel: N.M. Edinberry (*Sandwich, UK*), J. Emsley (*London, UK*), S.M. Hassur (*Washington DC, USA*), D.R. Kelly (*Cardiff, UK*), T. Laird (*Mayfield, UK*), T. Papenfuhs (*Frankfurt, Germany*), B. Pearson (*Wigan, UK*), J. Winfield (*Glasgow, UK*)

The chemical process industries are under increasing pressure to develop environmentally friendly products and processes, with the key being a reduction in waste. This timely new series will introduce different clean technology concepts to academics and industrialists, presenting current research and addressing problem-solving issues.

Feedstock Recycling of Plastic Wastes
by J. Aguado, *Rey Juan Carlos University, Móstoles, Spain*; D.P. Serrano, *Complutense University of Madrid, Spain*

Applications of Hydrogen Peroxide and Derivatives
by C.W. Jones, *formerly of Solvay Interox R & D, Widnes, UK*

Clean Synthesis Using Porous Inorganic Solid Catalysts and Supported Reagents
by J.H. Clark and C.N. Rhodes, *Clean Technology Centre, Department of Chemistry, University of York, UK*

How to obtain future titles on publication

A standing order plan is available for this series. A standing order will bring delivery of each new volume upon publication. For further information please contact:

Sales and Customer Care
Royal Society of Chemistry
Thomas Graham House
Science Park
Milton Road
Cambridge
CB4 0WF
Telephone: +44(0) 1223 420066

RSC
CLEAN TECHNOLOGY
MONOGRAPHS

Clean Synthesis Using Porous Inorganic Solid Catalysts and Supported Reagents

James H. Clark and Christopher N. Rhodes
Clean Technology Centre, Department of Chemistry, University of York, UK

ROYAL SOCIETY OF CHEMISTRY

ISBN 0-85404-526-0

A catalogue record for this book is available from the British Library

Published by The Royal Society of Chemistry,
Thomas Graham House, Science Park, Milton Road
Cambridge CB4 0WF, UK

For further information see our web site at www.rsc.org

Typeset by Paston PrePress Ltd, Beccles, Suffolk
Printed and bound by MPG Books Ltd, Bodmin, Cornwall

Preface

The chemical industry represents a highly successful sector of manufacturing and a vital part of the economy in many industrialised and developing countries. The range of chemical products is vast and these make an invaluable contribution to the quality of our lives. However, the manufacture of chemical products also leads to enormous quantities of environmentally harmful waste. The public image of the chemical industry has badly deteriorated in recent years due largely to concerns of adverse environmental impact, and public pressure and the work of action groups have played a major role in forcing action from the authorities on environmental issues. Increasingly demanding national and transnational (*e.g.* European) legislation is leading to a revolution in the chemical industry with the reduction or elimination of waste now being a central issue to the industry, the authorities and the general public. Industry is increasingly realising that high environmental standards are a lifeline to profitability in the highly competitive global and community markets that exist today. The so-called 'triple bottom line', which seeks simultaneous economic, environmental and societal benefit, is seen as a realistic evolutionary goal in chemical manufacturing.

National and international organisations have recognised the important contribution that cleaner processes and cleaner synthesis can make to environmental protection. In the early 1990s the United Nations Environmental Programme launched a number of industry sector working groups to coordinate and promote cleaner production technologies and practices. In Europe, the SUSTECH initiative was launched by the European Chemical Industry *via* CEFIC. This was aimed at promoting collaboration within the chemical and related processing industries on the theme of cleaner manufacturing. In the United States, the National Science Foundation and the Council for Chemical Research launched a programme called 'Environmentally Benign Chemical Synthesis and Processing' in 1992. In the United Kingdom, the Research Councils started a Clean Technology Programme in 1990, and by 1992 the 'Clean Synthesis of Effect Chemicals' initiative was running. Similar initiatives are now operating in countries around the world. The clean synthesis and processing initiatives have many similarities and many, if not all, include aspects of catalysis in the areas identified for support and encouragement.

Some of the major goals of waste minimisation are to enhance the intrinsic selectivity of any given process, to provide a means of recovering reagents in a form which allows easy recovery and regeneration, and to replace stoichiometric processes by catalytic ones. Solids, as catalysts or as supports for other reagents, offer potential for benefit in all of these areas. Unfortunately, most of the established routes to many fine and speciality chemicals and intermediates are based on liquid phase processes which either do not involve catalysts or use soluble catalysts which cannot be easily recovered. This means that organic chemists with the responsibilities for developing the commercial routes to such chemical products have little if any experience of working with solid catalysts or supports. The purpose of this monograph is to provide an overview of the properties of some of the more useful solid catalysts and supported reagents, and a survey of their most interesting and valuable applications in the preparation of organic chemicals in liquid phase reactions.

In Chapter 1, the principles of the fundamental subjects of waste minimisation, catalysis, adsorption, catalytic reactors and commercial heterogeneous catalytic processes are discussed. Solid catalysts offer many process engineering advantages compared to homogeneous processes including their non-corrosiveness, the wide range of temperatures and pressures that can be applied, and the easier separation of substrates and products from the catalyst. It is very important, however, to understand the important properties of solids in this context including porosity, surface characteristics including surface area and the dispersion of active sites. The mechanism of reactions employing solid catalysts is more complex than that of comparable homogeneous processes with the diffusion of substrate molecules to active sites and the diffusion of product molecules from the catalyst often being rate limiting. The physical form of the solid can be of vital importance and influences the choice of reactor. Solids can be used in all of the major types of reactor but either a particulate form or pelletised form of the solid will be required depending on the reactor. There are many established heterogeneous catalytic processes operating in industry, some on a very large scale. Apart from these, new processes are emerging often smaller in scale and where the main goal may be heterogenisation of the catalyst so as to improve reaction selectivity and catalyst lifetime and hence reduce waste.

In Chapter 2, the essential properties of zeolitic materials and some of their most interesting and potentially valuable applications in liquid phase organic reactions are considered. Zeolites are now well established in many very large scale petrochemical processes but have had much less impact in the fine and specialities chemicals areas. The essential properties of these materials – high thermal stability, easy recovery and reactivation, shape selectivity, and adjustable activity (giving them value in such diverse areas as acid catalysis and selective oxidations) – should make them useful in organic synthesis especially in the context of clean synthesis. The advent of mesoporous analogues further extends their value by enabling reactions to be carried out with larger substrates and products and through enhanced molecular diffusion rates. Some of the proven areas of application include ring hydroxylatlons, Friedel–Crafts acylations, Beckmann rearrangements, selective halogenations, and dehydration reactions.

Chapter 3 extends the coverage of the monograph to clay materials. Clays are readily available, inexpensive and with a longstanding reputation as versatile solid acid catalysts in large scale processes. More recently they have been shown to have a diverse range of uses as catalysts and catalyst supports in liquid phase organic reactions for the preparation of many useful chemical products. Some of the most important developments in the materials aspects of the subject are the use of acid-treated and ion-exchanged clays and the preparation of pillared clays which provide a more robust structure compared with the highly flexible natural layered clays. Their most promising applications include Diels–Alder reactions, Friedel–Crafts alkylations, hydrogenations and esterification reactions.

Chapter 4 is the largest in the book, which reflects the enormous level of current interest in the use of supported reagents as catalysts for liquid phase organic reactions of almost all types. The subject of supported reagents has matured from the original work on supporting stoichiometric reagents, so as to enhance activity through dispersion, to the heterogenisation of otherwise hazardous or in other ways difficult to use catalysts rendering them safe and easy to handle and recover, and in many cases, more selective in their chemistry. In this way, new environmentally benign processes based on hazardous catalysts such as aluminium chloride, boron trifluoride and sulfuric acid have been developed for reactions including Friedel–Crafts alkylations and acylations, and esterifications. The versatility of the concept is demonstrated by its successful application to base catalysis, oxidations and reductions, and to phase-transfer reactions. An understanding of the different methods of preparation of supported reagents and an appreciation of their relative advantages and disadvantages is very important. An increasing level of academic and industrial research activity in this area has led to the extension of the type of materials to chemically modified mesoporous solids. These offer the typical advantages of traditional supported reagents while offering better chemical and thermal stability. These advanced materials are already proving their value in areas including oxidation catalysis and various base-catalysed carbon–carbon bond forming reactions.

This monograph is not meant to be a comprehensive guide to the use of solid catalysts and supported reagents in the clean synthesis of organic chemicals. Many related subjects such as polymer supported reagents and metal oxides are beyond the scope of the book and are not covered in any length here although their importance is beyond question. The monograph does, however, seek to use important and varied examples of porous inorganic solid-catalysed organic reactions to illustrate the scope and potential of the subject. It also aims to provide fundamentally important information on heterogeneous catalysis and the preparation and use of solid catalysts in liquid phase organic reactions so as to assist the organic chemist inexperienced in this area to seek to exploit these exciting new process ideas. The Clean Technology revolution provides exciting opportunities for chemists and chemical engineers to develop new, safer, less wasteful and more environmentally acceptable chemical processes and products. Catalysis, with its established place at the heart of chemistry, is the ideal bedfellow for clean synthesis and we can look forward to an increasing number of cleaner catalytic processes in chemicals manufacturing.

Acknowledgements

We are indebted to May Price for her assistance in reconciling the problems of producing material from different computers and word-processing programmes and putting together the final form of the manuscript.

Contents

CHAPTER 1

Introduction

1 Waste Minimisation

Waste minimisation techniques can be grouped into four categories:

- Inventory management and improved operations
- Equipment modification
- Changes in the production processes
- Recovery, recycling and reuse

The waste minimisation approaches as largely developed by the Environmental Protection Agency (EPA) are given in Table 1.1. They can be applied across a wide range of industries including chemicals manufacturing.

2 Clean Synthesis

The hierarchy of waste management techniques has prevention as the most desirable option ahead of minimisation, recycling and, as the least desirable option, disposal. The term *cleaner production* embraces principles and goals that fall comfortably within the waste prevention–minimisation range. It has been described within the United Nations Environmental Programme as:

> *The continuous application of an integrated preventative environmental strategy to processes and products to reduce risks to humans and the environment. For production processes, cleaner production includes conserving raw materials and energy, eliminating toxic raw materials, and reducing the quantity and toxicity of all emissions and wastes before they leave a process.*

Cleaner processes fall under the umbrella of *waste reduction at source* and along with retrofitting, can be considered to be one of the two principal relevant technological changes. Waste reduction at source also covers good housekeeping, input material changes and product changes.[1] Within chemistry and the handling of chemicals the term *green chemistry* has become associated with

1

Table 1.1 *Waste minimisation approaches and techniques*

Approach	Techniques
Inventory management and improved operations	Inventory for all raw materials
	Use fewer toxic raw materials
	Produce fewer toxic chemicals
	Improvements in storage and handling
	Improve employee training
Equipment modification	Redesign production equipment so as to produce less waste
	Improve equipment operating efficiency
	Redesign equipment to aid recovery, recycling and reuse
Changes in the production process	Replace hazardous raw materials
	Optimise reactions
	Consider alternative low-waste routes
	Eliminate leaks and spills
	Consider product substitution
Recovery, recycling and reuse	Install closed-loop systems
	Recycle on site for reuse
	Properly segregate waste

the methods of waste reduction at source and more generally with reducing the environmental impact of chemicals and chemical processes.[2,3]

Within the context of cleaner production, terms such as *environmentally benign chemical synthesis* and *clean(er) synthesis* have often proven popular to help define the scope of national or trans-national programmes on waste minimisation. There is no widely accepted definition of clean synthesis but there is reasonable international agreement that the cleaner synthesis of chemicals, *i.e.* that involving a reduction in the toxicity and quantity of emissions and waste through changes to the process, is likely to be achieved through:[4]

- better use of catalysis
- alternative synthesis routes that avoid the need to use toxic solvents and feedstocks
- reduction in the number of synthetic steps
- elimination of the need to store or transport toxic intermediates or reagent
- novel energy efficient methods

It should be noted that catalysis features very highly on any list of preferred/ relevant technologies to help achieve a reduction in waste from chemical processes through the use of cleaner synthetic methods.

3 Catalysts and Catalysis

Catalysts are species that are capable of directing and accelerating thermo-dynamically feasible reactions while remaining unaltered at the end of the reaction. They cannot change the thermodynamic equilibrium of reactions.[5]

The performance of a catalyst is largely measured in terms of its effects on the reaction kinetics. The *catalytic activity* is a way of indicating the effect the catalyst has on the rate of reaction and can be expressed in terms of the rate of the catalytic reaction, the relative rate of a chemical reaction (*i.e.* in comparison to the rate of the uncatalysed reaction) or *via* another parameter, such as the temperature required to achieve a certain conversion after a particular time period under specified conditions. Catalysts may also be evaluated in terms of their effect on the *selectivity* of reaction, specifically on their ability to give one particular reaction product. In some cases, catalysts may be used primarily to give high reaction selectivity rather than high activity. *Stability* is another important catalyst property since catalysts can be expected to lose activity and selectivity with prolonged use. This then opens the way to *regenerability* which is a measure of the catalyst's ability to have its activity and/or selectivity restored through some regeneration process.

Catalytic processes are the application of catalysts in chemical reactions. In chemicals manufacture, catalysis is used to make an enormous range of products: heavy chemicals, commodity chemicals and fine chemicals. Catalytic processes are used throughout fuels processing, in petroleum refining, in synthesis gas ($CO + H_2$) conversion, and in coal conversion. More recently some aspect of clean technology or environment protection has driven most of the new developments. Many emission abatement processes are catalytic. An increasing number of catalytic processes employ *biocatalysis*. Most of these are fermentations classically carried out in stirred reactors using enzyme catalysts, which are present in living organisms such as yeast. Immobilised enzymes processes are becoming more common.

Catalysis is described as *homogeneous* when the catalyst is soluble in the reaction medium and *heterogeneous* when the catalyst exists in a phase distinctly different from the reaction phase of the reaction medium.

Almost all homogeneous catalytic processes are liquid phase and operate at moderate temperatures (< 150 °C) and pressures (< 20 atm). Corrosion of reaction vessels by catalyst solutions, and difficult and expensive separation processes are common problems. Traditionally the most commonly employed homogeneous catalysts are inexpensive mineral acids, notably H_2SO_4, and bases such as KOH in aqueous solution. The chemistry and the associated technology is well established and to a large extent well understood. Many other acidic catalysts such as $AlCl_3$ and BF_3 are widely used in commodity and fine chemicals manufacture *via* classical organic reactions such as esterifications, rearrangements, alkylations, acylations, hydrations, dehydrations and conden-sations. More recently there have been significant scientific and technological innovations through the use of organometallic catalysts.

Normally, heterogeneous catalysis involves a solid catalyst that is brought into contact with a gaseous phase or liquid phase reactant medium in which it is insoluble. This has led to the expression *contact catalysis* sometimes used as an alternative designation for heterogeneous catalysis. The situation can be rather more complicated with *phase transfer catalysis* (PTC) systems. Here the reactants themselves are present in mutually distinct phases, typically water and a non-aqueous phase (usually a hydrocarbon or halogenated hydrocarbon which has a very low solubility in water). The catalyst, which is normally a quaternary ammonium or phosphonium compound or a cation complexing agent such as a crown ether, is believed to operate at the interfacial region[6] and strictly need not be soluble in either the aqueous or non-aqueous phases. This is demonstrated by the activity of immobilised onium compounds (see Chapter 4). In practice, simple onium compounds such as tetraarylphosphonium compounds, which are insoluble in hydrocarbons, are inactive in corresponding hydrocarbon–water PTC systems, presumably because the low surface area of the salt provides little effective interfacial area for the catalysis to occur.

4   Heterogeneous Catalysts

Most of the large-scale catalytic processes take place with gaseous substrates contacting solid catalysts. The engineering advantages of these processes compared to homogeneous processes are:

- solid catalyst are rarely corrosive
- a very wide range of temperatures and pressures can be applied to suit the process and the plant (strongly exothermic and endothermic reactions are routinely carried out using solid catalysts)
- separation of substrates and products from catalysts is easy and inexpensive

Many solid catalysts are based on porous inorganic solids. The important physical properties of these materials are surface area (often very large and measured in hundreds of square metres per gram), pore volume, pore size distribution (which can be very narrow or very broad), the size and shape of the particles and their strength. The solid catalyst provides a surface, usually largely internal, for the substrates to adsorb and react on. Thus the surface characteristics of the surface (roughness, functional groups, organophilicity, hydrophobicity, *etc.*) are also vital to performance.

Typical heterogeneous catalysts used in large-scale industrial processes are complex materials in terms of composition and structure. Catalytically active phases, supports, binders and promoters are common components. They typically are activated in some way before use, often by calcination. Heterogeneous catalysts have been prepared for many years and often the preparation procedure used in industry is based more on operator experience and tradition than on sound science. Generally the support is prepared or activated before use with the actual catalytic species and any promoters are added later, often as aqueous solutions of precursor compounds, which are then converted into their final active forms by a final treatment step (*e.g.* calcination). The active sites in heterogeneous catalysts are often metal centres. At the surface these can be very different to those in the bulk, due to differences in ligand environment and coordination geometry. Generally metal surfaces offer the advantage over metal complexes of higher thermal stabilities. Supported palladium, for example, has largely replaced soluble palladium compounds in the manufacture of vinyl acetates.

Metal oxides are widely used as catalyst supports but can also be catalytically active and useful in their own right. Alumina, for example, is used to manufacture ethene from ethanol by dehydration. Very many mixed metal oxide catalysts are now used in commercial processes. The best understood and most interesting of these are zeolites that offer the particular advantage of shape selectivity resulting from their narrow microporous pore structure. Zeolites are now used in a number of large-scale catalytic processes. Their use in fine chemical synthesis is discussed in Chapter 2.

5 Heterogeneous Catalysis

The catalytic mechanism of reaction on solids can be broken down into five consecutive steps:

1. Substrate diffusion
2. Substrate adsorption
3. Surface reaction
4. Product desorption
5. Product diffusion

Substrate molecules must diffuse through the network of pores to reach the internal region and the product molecules must diffuse out of the pore network. Smaller pores provide the advantage of large surface areas and high particle mechanical strengths but lead to problems with slow molecular diffusion. This can lead to concentration gradients where the substrate concentration is at a maximum at the external surface of the particle while the product concentration is at a maximum at the centre of the particle. Large concentration gradients will mean poor catalyst effectiveness.

In the case of a solid catalyst operating in a liquid phase reaction system the problems of diffusion and concentration gradients can be particularly severe. Substrate diffusion can be further broken down into two steps, external diffusion and internal diffusion. The former is controlled by the flow of substrate molecules through the layer of molecules surrounding catalyst particles and is proportional to the concentration gradient in the bulk liquid, *i.e.* the difference in the concentrations of the substrate in the bulk medium and at the catalyst surface. The thickness of the external layer in a liquid medium is dependent on the flowing fluid and on the agitation within the reaction system; typically it is 0.1–0.01 mm thick. Internal diffusion of substrate molecules is a complex process determined not only by the resistance to flow due to the medium but also by the constraints imposed by the pore structure. As stated earlier, the latter is especially important with microporous solids, *i.e.* when the pore geometries are comparable to molecular geometries. Diffusional limitation, be it due to external, or more commonly, internal, resistance to motion means that the actual (observed) rate of reaction will always be lower than that predicted on the basis of the intrinsic activity of the available surface of the catalyst. Furthermore, the actual rate of reaction can never be faster than the maximum rate of diffusion of the substrate molecules. Apart from mass transfer considerations, heat transfer also becomes of considerable importance in commercial scale processes. Since reaction is either endothermic or exothermic, and reaction occurs at the (internal or external) surface of the catalyst, a temperature gradient will be established between the catalyst particle surface and the external medium. This will depend on the heat of reaction, the activity of the catalyst and the thermal properties of the solid and other phases. Since temperature affects the rate of reaction, heat transfer calculations can become extremely complex and the data that are calculated can be unreliable.

6   Adsorption by Powders and Porous Solids

Adsorption is the enrichment of material or increase in the density of the fluid close to an interface. Under certain conditions this results in an appreciable enhancement in the concentration of a particular component which is dependent on the surface or interfacial area. Thus all industrial adsorbents and the majority of industrial heterogeneous catalysts have large surface areas of > 100 $m^2 g^{-1}$ based on porous solids and/or highly particulate materials.[7] In the simplest case for spherical particles of density r and all of diameter d, the specific surface area s, can be defined as:

$$s = 6/rd$$

Thus for a powder made up of smooth particles of diameter 10^{-6} m and density $2 \, g \, cm^{-3}$, the specific surface area would be $3 \, m^2 \, g^{-1}$. In reality powder particles are irregular and are clustered together in aggregates. These aggregates may be broken down by grinding. The aggregate can itself be regarded as a secondary particle, which contains some internal surface often larger than the external surface. Thus the aggregate possesses a pore structure. The size of the pores in porous solids can be classified as micro, meso or macro based on their width as measured by some defined method. It is often difficult to distinguish between porosity and roughness or between pores and voids, although a useful distinction is to reserve porosity for materials with irregularities deeper than they are wide.

Adsorption is brought about by the interactions between the solid and the molecules in the fluid phase. The forces involved are classified as chemisorption (chemical bonding) or physisorption (non-chemical bonding). Some of the main distinguishing features are:

- physisorbed molecules keep their identities and desorb back to the fluid

phase unchanged, whereas chemisorbed molecules can be expected to change as a result of adsorption and are not recovered unchanged on desorption
- chemisorption is generally restricted to a monolayer whereas at high enough pressures, physisorption can produce multilayers
- physisorption is exothermic (commonly tens of kilojoules per mole) but tends to involve energies below those typical of chemical bond formation, whereas chemisorption involves energies of the same magnitude as chemical bond formation

Some of the principal terms and properties of adsorption, powders and porous solids are given in Table 1.2.

7 Reactor Types

Solid catalysts can be used in all of the major reactor types, *batch, semibatch, continuous stirred tank* and *tubular*. In the first three cases particulate (powder) catalysts would be appropriate, whereas with the tubular reactor the catalyst would often need to be formed into pellets.[8,9]

Batch reactors using particulate catalysts need to be well stirred in order to give uniform compositions and to minimise mass transport limitations. They are likely to be preferred for small-scale production of high-priced products or

Table 1.2 *Definitions associated with adsorption, powders and porous solids*

Term	Definition
Adsorption	Enrichment in an interfacial layer
Adsorbate	Substance in the adsorbed state
Adsorbent	Solid material on which adsorption occurs
Adsorption isotherm	The relation at constant temperature between the amount adsorbed and equilibrium pressure or concentration
Chemisorption	Adsorption involving chemical bonding
Physisorption	Adsorption without chemical bonding
Monolayer	Amount required to cover the entire surface
Powder	Discrete particulate material (particle dimension < *ca.* 1 mm)
Surface area	Available surface as defined by a particular method
External surface area	Area of surface outside of pores
Internal surface area	Area of pore walls
Porous solid	Solid with cavities or channels which are deeper than they are wide
Void	Space between particles
Micropore	Pore of internal width of < 2 nm
Mesopore	Pore of internal width of 2–50 nm
Macropore	Pore of internal width of > 50 nm
Pore size	Pore width
Pore volume	Volume of pores (defined by stated method)
Porosity	Ratio of total pore volume to apparent volume of particle

when continuous flow is difficult. The separation of the catalyst from the organic components in a batch reactor may not be simple. If the particles settle well, then the liquid can be removed by decantation and the vessel can be subsequently recharged with fresh substrate(s). Otherwise, it may be necessary to separate *via* filtration or centrifugation, which requires additional equipment and adds to process time. Batch reactors are commonly used in fine/speciality chemicals manufacturing companies and it is important that solid catalysts can be amenable to such reactor configurations so as to make the catalyst technology accessible and attractive to these companies. Smaller and more specialised companies are unlikely to be prepared to invest in new equipment so as to exploit new chemistry unless the whole technology is clearly proven and there is a secure long-term profitable market for the products.

The semibatch reactor with the continuous addition or removal of one or more of the components offers an added degree of sophistication, which can benefit the process through greater stability and safer operation. This method may also lend itself to liquid–particulate solid reactions where a bulk substrate is continuously being converted over a catalyst into a product. For example, in aerial oxidations of substrates, continuous removal of the reaction mixture (through a suitable frit to prevent transfer of solid catalyst) followed by recycling of the unreacted (lower boiling) substrate will enable large total amounts of product to be produced from one catalyst batch and in one reactor.

The continuous stirred tank reactor (CSTR) adds a further degree of sophistication and is generally preferred to single batch operations for the larger scale or more frequent manufacture of products due to lower operating costs and overall investment. In practice, mechanical or hydraulic agitation is required to achieve uniform composition and temperature.

The tubular reactor is a vessel through which the flow is continuous. There are several configurations of tubular reactors suitable for multiphase work, *e.g.* for liquid–solid and gas–liquid–solid compositions. The flow patterns in these systems are complex. A fixed bed reactor is packed with catalyst, typically formed into pellets of some shape, and if the feed is single phase, a simple tubular plug-flow reactor may suffice (Figure 1.1). Mixed component feeds can be handled in modifications to this.

The moving bed reactor can be used when catalyst deactivation is a major factor (*i.e.* when the lifetime of the fixed bed catalyst is low); here spent catalyst is slowly removed from the reactor while fresh material is slowly added at the top (Figure 1.2).

Low feed rates are suitable for trickle bed reactors where for gas–liquid–solid mixing, the gas and the liquid are fed into the top of the reactor. This gives long gas residence times but short liquid residence times. Such a configuration is often used in hydrogenation reactions. When the gas–liquid is fed into the bottom of the reactor, it is known as a bubble reactor. Here the gas residence times are short but the liquid residence times are relatively long. This is commonly used in oxidation reactions. Heat transfer can be a major problem with both trickle and bubble reactors and in such cases a slurry bubble column reactor can be employed.

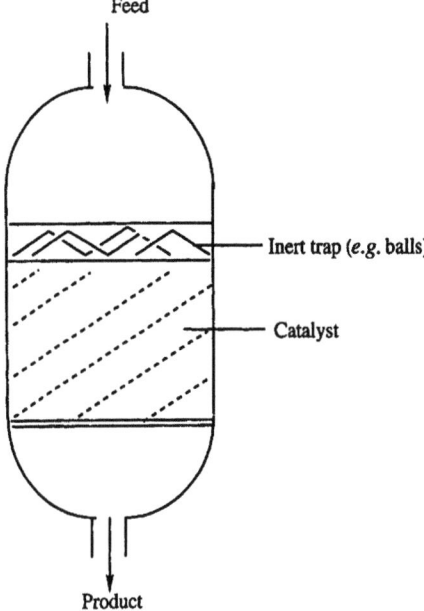

Figure 1.1 *Fixed bed reaction (adiabatic)*

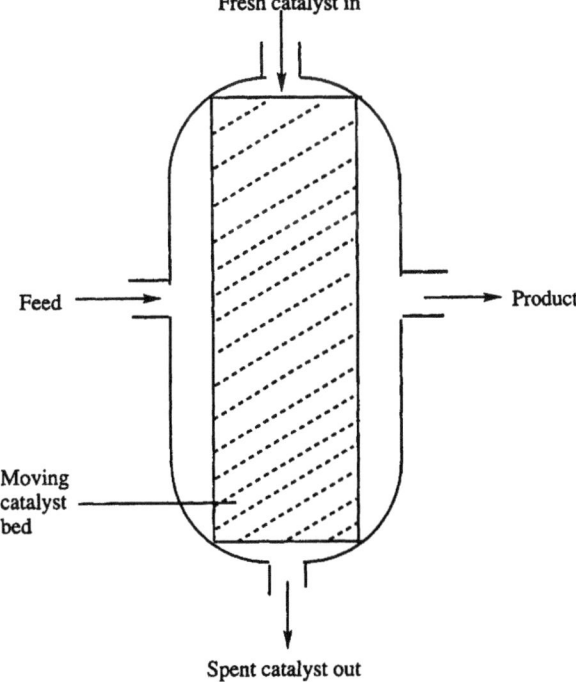

Figure 1.2 *Moving (radial) fixed-bed reactor*

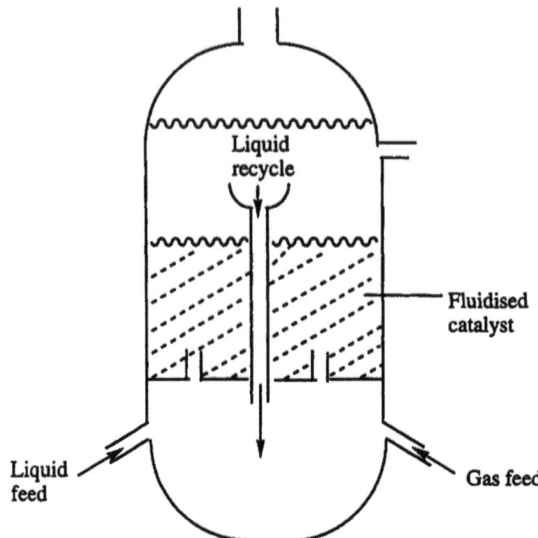

Figure 1.3 *Gas–liquid–solid fluidised reactor*

It is possible to use solid catalysts in particulate forms in tubular reactors through the use of fluidised or fluid bed reactors, where the upward flow of the feed is sufficient to suspend the particulate catalyst in such a way that it seems to behave like a liquid (Figure 1.3). It is however preferable to use more structured catalysts, since better flow characteristics can be achieved, thus minimising hydrodynamic uncertainties and maximising volumetric reaction rates.

8 Commercial Heterogeneous Catalytic Processes

Catalysts played a major role in establishing the economic strength of the chemical and related industries in the first half of the 20th century and an estimated 90% of all of the chemical processes introduced since 1930 depend on catalysis. This has resulted in the build up of an enormous worldwide market for catalysts, which is valued today at some $5000 million per annum with the product value dependent upon them being a staggering $250 000 million.

Heterogeneous catalysis is especially important in industry. Some of the major industrial processes that use solid catalysts include the synthesis of inorganic chemicals such as NH_3, SO_3 and NO, the various reactions used in the refining of crude petroleum such as cracking, isomerisation and reforming, and many of the major reactions of the petrochemical industry, such as the synthesis of methanol, the hydrogenation of aromatics and various controlled oxidations. Some of the major industrial processes to be catalysed by inorganic solids are shown in Table 1.3.

In the long-established manufacturing process of ammonia, for example, 100 megatonnes of ammonia requires some 40 megatonnes of hydrocarbons, 85 megatonnes of water and 80 megatonnes of nitrogen from the air, through 7–8

Table 1.3 *Some large-scale processes catalysed by inorganic solids*

Catalyst	Process
Mixed iron and molybdenum oxides	$CH_3OH + O_2$ (HCHO)
Solid acids (*e.g.* zeolites)	Paraffin cracking and isomerisation; alkylation; olefin polymerisation
Mixed metal oxides	$CH_3CH{=}CH_2$ (acrolein)
Alumina	ROH (olefin + H_2O)
Supported metals	$H_2 + N_2$ (NH_3)
Supported tungsten or rhenium	Olefin metathesis
Metals (Ni, Pd, Pt) and supported metals	C=C bond hydrogenation
Metals (Cu, Ni, Pt)	C=O bond hydrogenation

successive process units, of which only one, the adsorption of CO_2, does not involve heterogeneous catalysis. Over 80% of the components of crude oil processed come into contact with heterogeneous catalyst within the refineries. All of the unit syntheses used in the manufacture of methanol use heterogeneous catalysis.

Heterogeneous catalysis is playing an increasingly important role in smaller scale chemical manufacturing, often with the result of a major reduction in waste. A good example of this is the new heterogeneous route to hydroquinone based on the titanium silicate catalyst TS-1. The traditional homogeneous route involved the oxidation of aniline with manganese dioxide and sulfuric acid followed by reduction with Fe/HCl. This route led to very large volumes of hazardous waste (four mole equivalents of manganese sulfate are produced, for example). The new route is an excellent example of clean synthesis. The catalyst is reusable, the oxidant is relatively safe to handle and the by-products are either innocuous (water) or marketable (catechol) (Figure 1.4).

Supported catalysts are extremely useful in almost all areas of petroleum refining and commodity chemical processing. As a group they are the major contributor to the catalyst industry, with about a third of the market being for petroleum refining, a third for chemicals processing and a third for emission control. They offer significant advantages on the large-scale plant, notably reduced cost (compared to using the unsupported catalyst), easy separability

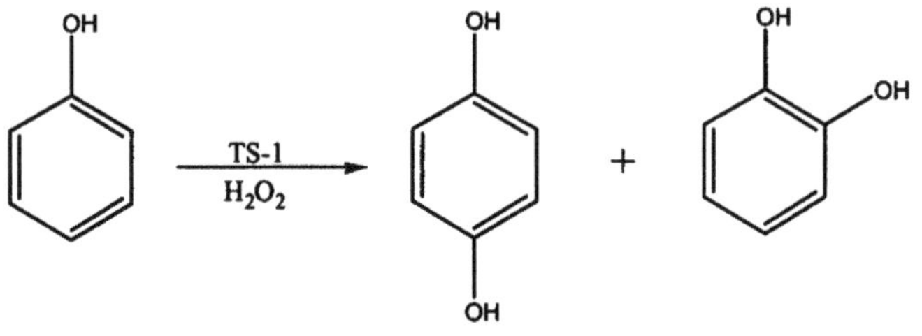

Figure 1.4 *New heterogeneous catalytic route to hydroquinone*

and improved activity and selectivity. Recent innovations in the applications of supported catalysts have included catalytic distillations, the use of catalytic membranes and the widespread use of modern automotive catalytic converters. It is also expected that heterogeneous catalysts, including supported catalysts, will play an increasingly important role in the manufacture of fine chemicals.[10]

9   Role of Catalysis in Industrial Waste Minimisation

Remarkably, the catalysis market continues to grow and the market potential is considered to be very large. This is partly due to the rapid growth in the use of catalysts to control the emission of pollutants, most famously in automobile exhaust catalysis, which accounted for about one third of the total US catalyst market in the 1990s. Another major growth area is likely to be in pollution prevention and waste minimisation through the introduction of catalysts into processes where catalysis has not previously been used and through the introduction of improved catalysts which give improved product quality or process efficiency and reduced waste. Despite the pre-eminence of catalysis in large-scale continuous petrochemical processes such as cracking, isomerisation and alkylation, their use in smaller scale continuous or batch-type processes is far from common.

The crucial factor is the introduction of *enviro-economics* as a driving force for new products and processes including new catalysts and catalytic processes. Owing to increasing environmental pressures and the subsequent increase in environmental legislation over the last ten years, industry now has to meet the added costs of cleaning up its act or risk being put out of business. Remarkably it seems that only as a result of the activities of environmentalists in the 1980s and of the regulatory authorities in the 1990s is industry now waking up to the fact that the basic requirements of reduced costs, improved public image and compliance with environmental law can be met through waste minimisation strategies.[11] The obstacle to the introduction of new cleaner technology has been capital costs. While capital costs will always be an important issue, increasing global competition, more demanding environmental legislation, an increasing emphasis on lower volume, higher value products and hopefully a more buoyant world economy should ensure the widespread introduction of cleaner process technologies.

While the use of catalysts in secondary pollution prevention, *i.e.* the clean up of waste, has become well established and is likely to grow well into the 21st century, it is in primary pollution prevention (pollution reduction or avoidance) where there should be a spectacular growth in application and importance. The key areas of clean technology where catalysis can have a major impact are:

- elimination of toxic reagents and intermediates
- increases in plant utilisation and a reduction in the number of process steps
- reduction in toxic emissions and waste streams

Catalysis using solid catalysts is rapidly emerging as a new *enviro-technology* designed to enhance process efficiency and reduce process waste through more efficient use of plant, lower energy costs and reduced side-products or to replace or remove the need for environmentally unacceptable hazardous reagents, intermediates and catalysts.[12,13] There are several significant examples of new industrial processes based on these concepts. At the very large scale end, the ethylation of benzene *en route* to styrene is now largely carried out using a zeolite catalyst which replaces the hazardous alkylation catalysts hydrogen fluoride and aluminium chloride. The use of zeolites as catalysts in typically smaller scale, liquid phase chemical reactions is described in Chapter 2.

The porous titanium silicate TS-1 represents one of the great commercial successes of recent years. Despite only being reported for the first time in the last decade, it is already established as an oxidation catalyst in the manufacture of hydroquinone, and processes based on its use as a catalyst in the epoxidation of propene and the ammoxidation of cyclohexanone are near the production stage.[14] The use of the increasingly diverse range of molecular sieve solid catalysts is also described in Chapter 2.

Clays, which have themselves proven popular solid catalysts for many years, form the basis of new commercial supported reagent catalysts developed for the liquid-phase synthesis of fine chemicals.[15] Particularly significant is their use in Friedel–Crafts reactions, which represent a remarkably diverse and frequently employed class of organic reactions used in the manufacture of countless intermediates and products. Here heterogeneous catalysis is a relative new-comer, since most reactions are carried out in batch-type reactors rather than in continuous fixed-bed reactors where solid catalysts are so commonly employed. The essential logic behind the use of catalysts and supports such as acid-treated clays in these reactions is that their mesoporous nature makes them more likely candidates for many liquid phase reactions. The more microporous zeolitic materials are often less suitable because of poor molecular diffusion rates in the liquid phase, especially when the molecules are quite large or polar. The new solid catalysts are meant to replace existing reagents and catalysts, which are environmentally unacceptable. Most notoriously, aluminium chloride, perhaps the most widely used Friedel–Crafts catalyst (at least in batch processes) is the source of enormous quantities of toxic waste. In Friedel–Crafts acylations for example, greater than stoichiometric quantities of aluminium chloride are normally used as a result of the complexation of the Lewis acid by the product (Lewis base) ketone on a molecule-by-molecule basis. Reaction leads to the production of a complex which is routinely broken down by a water quench leading to the evolution of large volumes of hydrogen chloride gas (toxic emissions) and the production of a toxic waste stream made up of water, aluminum salts, acid and trace organics. This is a good example of the type of chemical process that is unlikely to be environmentally acceptable in the future, and where the costs of clean-up and waste disposal will make it difficult to maintain economically. The use of clays as catalysts is described in Chapter 3, while supported reagents are described in Chapter 4.

10   Heterogenisation

Apart from the use of the now well established microporous zeolitic solids as catalysts and the emerging use of mesoporous solids as catalysts, there is also a growing interest in the related area of heterogenisation. Here, an active compound or complex is immobilised through binding to an insoluble solid. The solid is commonly a mesoporous solid so that the useful properties of the solid support (high surface area, high concentrations of active sites within pores) can be combined with the activity of the compound or complex. Alternatively, the catalyst can be an insoluble cross-linked polymer.[16] While enhancement in catalytic performance, be it in terms of activity and/or selectivity, is clearly desirable, the principal motive for heterogenisation is to facilitate separation, recovery and reuse. Easier handling and lower toxicity can also be achieved through heterogenisation. The earliest examples of so-called supported reagents were also aimed at overcoming very low reagent activity due to low surface areas, high lattice energies and low solubilities. Non-catalytic materials such as KMnO$_4$–silica, NaSCN–alumina (*i.e.* those where the reagent was spent on use and could only be reused after a separate regeneration stage) and catalytic materials such as ZnCl$_2$–montmorillonite (*i.e.* those where the reagent is not chemically changed on use and could possibly be reused after a reactivation stage) relied on physisorption to keep the support and active species together.[17–19] The disadvantage of these loosely bonded materials is obvious and partial destruction of the materials with leaching into the reaction solution or during separation and work-up are serious problems (interestingly some of the more valuable of these materials such as the supported fluorides turn out to be more complex and stable than many others due to reaction between the support and the active species giving robust chemisorbed active sites). Some examples of well established supported reagents are given in Table 1.4.

In more recent years, attention has at least partly switched to the development of heterogenised compounds and complexes where the active sites are

Table 1.4 *Some well established supported reagents*

Supported reagent	Applications
KF–alumina and other supported fluorides	Various base-catalysed reactions
KF–CaF$_2$	Nucleophilic fluorinations
KMnO$_4$–silica, *etc.*	Oxidations, including RCH$_2$OH → RCO$_2$H
KCN–alumina, *etc.*	Nucleophilic cyanations
KSCN–alumina, *etc.*	Nucleophilic thiocyanations
ZnCl$_2$–clay (K10)	Friedel–Crafts alkylations
Fe(NO$_3$)$_3$–clay (K10)	Nitrations and oxidations
KOH–alumina	Various base-catalysed reactions
t-BuOCl–zeolite	*para*-Selective aromatic monochlorinations
K$_2$Cr$_2$O$_7$–alumina	Various oxidations
NaBH$_4$–silica	Various reductions

chemically bonded to the support. The immediate advantages of better stability and lower tendency to leach, which can also greatly facilitate reuse of the material, must be balanced with increased complexity in material synthesis and the fact that a compound or complex that is chemically immobilised onto a support material cannot be considered to be an exact equivalent of the 'free' analogue (typically in solution). A greater similarity between the immobilised and free species can be achieved through the use of substantial spacer groups between the support and the active centre. In this way, at least some of the more direct effects of the support can be 'distanced' from the reaction zone and made less significant. If it is desirable for the the immobilised species to behave as similarly as possible to the free analogue, then it is also important to maintain local structural integrity around the active centres; spacer groups and support-species bridging groups should be distant from the active centre.

In some cases the heterogeneous version of a catalyst can be prepared by direct reaction of that catalyst with a suitable support material. Thus reactive Lewis acids such as aluminium chloride will react with hydroxylated materials such as silica gel to give directly bonded surface species such as $-OAlCl_2$.[20] Another single-step route to the supported catalyst is *via* sol-gel techniques, typically to produce an organically modified mesoporous silica. This is based on the co-polymerisation of a silica precursor and an organosilicate precursor (Figure 1.5).

More commonly, however, the heterogenised versions of catalysts are prepared by multi-stage routes. These include the grafting of silanes (or possibly other reactive reagents that possess appropriate functionality) onto a support material. The catalytic group can be present in the silane, which is attached to the surface, or more commonly can be introduced by post-modification reactions. The latter is usually necessary because of the limited range of silanes available. The inexpensive 3-aminopropyl(trimethoxy)silane is a popular choice, since it behaves like a typical amine function and can be derivatised by formation of amides or imines and by alkylation. Drawbacks with this approach include the formation of several surface species resulting from the binding of one, two or three Si–O–Si groups, attachment of oligomeric silanes, and the presence of physisorbed species. Another less frequently used method is surface chlorination followed by reaction of the Si–Cl groups with an organo-metallic compound such as a Grignard reagent. This has the advantage over the other methods of forming a direct Si–C bond at the surface (which is relatively

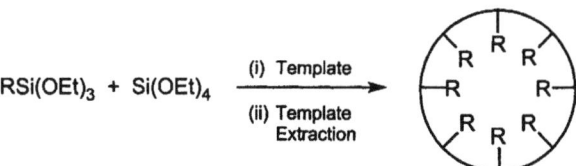

Figure 1.5 *Preparation of an organically modified mesosporous silica via sol-gel methodology*

stable) and precludes the formation of surface bound oligomers and variable modes of attachment.[12,20]

Methods for the introduction of reactive groups onto organic polymers follow similar lines.[16] Thus a pre-formed support can be chemically modified in a single, or more often, multi-step procedure. Alternatively, the reactive group can be introduced during resin preparation by using a conventional co-monomer already carrying the reactive group required.

Methods of heterogenisation, examples of the catalysts that have been successfully prepared and their use in catalysis are discussed in Chapter 4.

References

1 T. Lester, in 'Chemistry of Waste Minimisation', ed. J.H. Clark, Blackie Academic, London, 1995, Chapter 1.
2 'Green Chemistry: Challenging Perspectives', eds. P. Tundo and P.T. Anastas, Oxford Science, Oxford, 1999.
3 P.T. Anastas and J.C. Warner, 'Green Chemistry: Theory and Practice', Oxford University Press, Oxford, 1998.
4 M. Braithwaite, in 'Chemistry of Waste Minimisation', ed. J.H. Clark, Blackie Academic, London, 1995, Chapter 2.
5 B.C. Gates, 'Catalytic Chemistry', John Wiley, New York, 1992.
6 Y. Goldberg, 'Phase Transfer Catalysis: Selected Problems and Applications', Gordon and Breach Science Publishers, Yverdon, Switzerland, 1992.
7 F. Rouquerol, J. Rouquerol and K. Singh, 'Adsorption by Powders and Porous Solids', Academic Press, San Diego, 1999.
8 K.R. Westerterp, W.P.M. van Swaaji and A.A.C.M. Beenackers, 'Chemical Reactor Design and Operation', John Wiley, New York, 1984.
9 H.S. Fogler, 'Elements of Chemical Reactor Engineering', 2nd Edn., P.T.R Prentice Hall, Englewood Cliffs, NJ, 1992.
10 'Heterogeneous Catalysis and Fine Chemicals IV, Studies in Surface Science and Catalysis', Vol. 108, Elsevier, Amsterdam, 1997.
11 'Waste Minimisation: A Chemist's Approach', ed. K. Martin and T.W. Bastock, Royal Society of Chemistry, Cambridge, 1994.
12 J.H. Clark, 'Catalysis of Organic Reactions Using Supported Inorganic Reagents', VCH, New York, 1994.
13 J.H. Clark, *Green Chemistry,* 1999, 1.
14 J.H. Clark and D.J. Macquarrie, *Org. Process Res. Dev.*, 1997, **1**, 413.
15 T.W. Bastock and J.H. Clark, in 'Speciality Chemicals', ed. B. Pearson, Elsevier, London, 1992.
16 D.C. Sherrington, in 'Chemistry of Waste Minimisation', ed. J.H. Clark, Blackie Academic, London, 1995, Chapter 6.
17 J.H. Clark, A.P. Kybett and D.J. Macquarrie, 'Supported Reagents: Preparation, Analysis and Applications', VCH, New York, 1992.
18 'Preparative Chemistry using Supported Reagents', ed. P. Laszlo, Academic, San Diego, 1987.
19 'Solid Supports and Catalysts in Organic Synthesis', ed. K. Smith, Ellis Horwood, Chichester, 1992.
20 J.H. Clark and D.J. Macquarrie, *Chem. Commun.*, 1998, 853.

CHAPTER 2

Zeolitic Materials

1 Introduction

The term zeolite was coined in 1756 by the Swedish mineralogist Cronstedt who observed that the mineral stilbite frothed and gave off steam when heated.[1] The name zeolite comes from the Greek meaning 'boiling stone'. Zeolites have a crystal structure which is constructed from TO_4 tetrahedra, where T is either Si or Al. In addition to the relatively small number of naturally occurring zeolites there is a wide range of synthetic materials.[2]

The large number of synthetic zeolites has led to a complicated naming system. All of these materials can, however, be described in terms of structure types that define how the tetrahedra are linked together.[3] Each structure type is given a unique framework code as shown in Table 2.1. For example, zeolite A (also called 3A, 4A and 5A depending on the type of exchangeable cation) all share the LTA framework.

The size of the aperture which controls entry into the internal pore volume is determined by the number of T atoms and oxygens in the ring. The apertures are classed as ultralarge (> 12 membered ring), large (12) medium (10) or small (8). Aperture sizes range from 0.4 nm for 8 ring structures such as zeolite A, 0.54 nm for 10 rings such as ZSM-5, to 7.4 nm for 12 rings such as zeolite X and ZSM-12.

It is possible to fine tune the pore-opening of a zeolite to allow the adsorption of specific molecules. One method is to change the exchangeable cation. For

Table 2.1 *Zeolite codes and ring sizes*

Zeolite	Framework code	Number of tetrahedra in ring
Sodalite	SOD	4
Zeolite A	LTA	8
Erionite-A	ERI	8
ZSM-5	MFI	10
Faujasite	FAU	12
Mordenite	MOR	12 and 8
Zeolite-L	LTL	12

example, when Na^+ ions are replaced by Ca^{2+} ions in zeolite A, the effective aperture increases. The other method used for tuning the pore openings is to change the Si/Al ratio. An increase in the ratio of Si to Al will (i) slightly decrease the unit cell size, (ii) decrease the number of exchangeable cations, thus freeing the channels, and (iii) make the zeolite more hydrophobic in character.

2   Compositions

The general formula for aluminosilicate zeolites is:

$$x M_{2/n}O \cdot x Al_2O_3 \cdot y SiO_2 \cdot z H_2O$$

The framework carries a net negative charge equal to the number of tetrahedral aluminium ions. The negative charge is balanced by a corresponding number of non-framework cations, M. The non-framework cations are usually sited in, or have access to, the pores and can be readily be exchanged for other ions by treatment with a suitable salt solution. The last component is the aqueous sorbed phase, which can be removed from the sample, without any change to the aluminosilicate framework, by heat treatment. These three components; the framework, the non-framework cations and the sorbed phase, can each play an important role in determining the catalytic properties of zeolites.[4]

3   Synthesis

The hydrothermal conditions in geology which give rise to natural zeolites can be reproduced in the laboratory. Barrer demonstrated in the 1940s that a series of zeolites could be synthesised under hydrothermal conditions.[5] The synthesis procedure involves the mixing of a soluble alumina component, a silica component and an inorganic base. This mixture forms a gel which is allowed to crystallize under autogenous pressure, for a period of between a few hours to several weeks at temperatures between 40 and 200 °C. The optimum crystallisation time is determined by taking a series of samples over time and analysing these by powder X-ray diffraction. Crystallisation is complete when the peaks in the X-ray diffraction pattern have reached a maximum in intensity. Varying the inorganic base in the synthesis procedure gives rise to a range of zeolite structures.

4   Zeolite Catalysis

Synthetic zeolites were developed for fluid catalytic cracking in the early 1960s.[6] This process occurs *via* carbonium ion intermediates and is therefore catalysed by Brönsted acids. These sites are normally protons attached to bridging framework oxygen atoms and are introduced into the zeolite *via* ion exchange. For example, exchange of sodium zeolite Y with ammonium ions gives the NH_4^+ form, which on heating loses NH_3 to leave the proton exchanged zeolite (Reaction 1).

Zeolite as synthesised

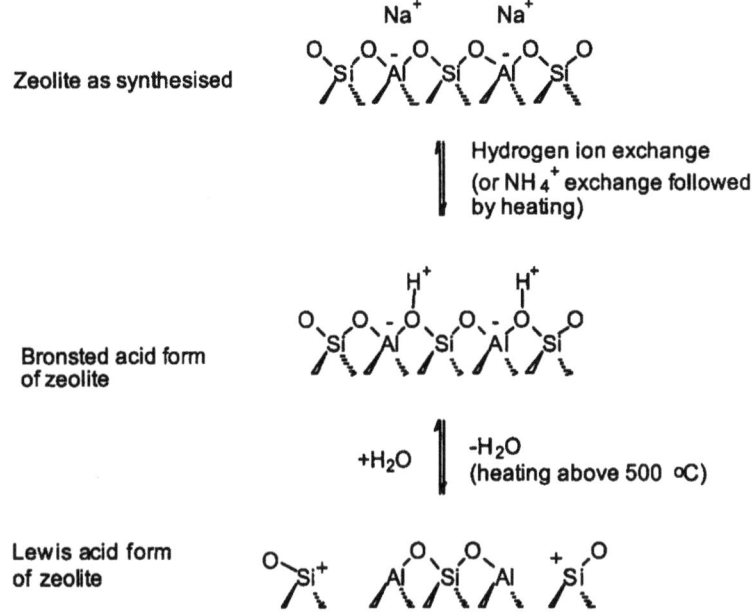

Hydrogen ion exchange
(or NH_4^+ exchange followed
by heating)

Bronsted acid form
of zeolite

$+H_2O$ $-H_2O$
(heating above 500 °C)

Lewis acid form
of zeolite

Figure 2.1   *Generation of Lewis and Brönsted acid sites in zeolites*

$$NH_4^+ \{zeolite\} \rightarrow NH_3(g) + H^+ \{zeolite\} \tag{1}$$

Further heating removes water from the Brönsted acid site, exposing a three
fold coordinated Al ion which has Lewis acid character. A reaction scheme for
the formation of these sites is shown in Figure 2.1

The surfaces of zeolites can thus display either Brönsted or Lewis acid sites or
a combination of the two depending on how the zeolite is prepared. Brönsted
acid sites are converted into Lewis acid sites as the temperature is increased
above approximately 500 °C and water is driven off. The strength of the acid
sites is directly related to the framework composition of the zeolite. Zeolites
with a high Si:Al ratio have the strongest acid sites.[7]

A special feature of zeolites which makes them such superb catalysts in some
cases is their shape selectivity. The shape selectivity may arise in three ways:
reactant selectivity, product selectivity and, of lesser importance, transition
state selectivity. Reactant selectivity arises from the ability of only certain
molecules to be absorbed into the zeolite cavities and thus reach the active acid
sites. An important commercial process that exploits this type of reactant
selectivity is catalytic dewaxing. Compared to the branched isomers, the straight
chain alkanes have low octane numbers and contribute to wax formation in
diesel fuel. Product selectivity is derived from the fact that only certain products
are of the correct dimension to escape from the zeolite once they have been
formed. Transition state selectivity relies upon the fact that certain intermediates, which are formed during a chemical reaction at the active site, will not fit in

the cavity; such a reaction is barred from occurring and the reaction will proceed along a different route to a different product.

5 Isomorphously Substituted Zeotypes

Other framework structures containing atoms such as aluminium and phosphorus tetrahedrally coordinated by oxygen have been synthesised and are given the generic name zeotypes. Pure aluminium phosphate (commonly known as ALPO) and its derivatives have been found to take the same structural forms as some of the zeolites, such as sodalite and faujasite, as well as some novel structures. The metal–aluminium phosphates can be formed with metals such as Li, Be, Mg, Mn, Fe and Zn replacing some of the aluminium and these are called MeALPOs. If the compound contains silicon or silicon and a metal, partially replacing aluminium or phosphorus leads to SAPOs and MeSAPOs.

In the same way that replacement of Si^{4+} by Al^{3+} in zeolite structures leads to the formation of Brönsted acid sites, so does the replacement of Al^{3+} by divalent metals in ALPOs. Consequently there has been much recent interest in these materials as potential heterogenous catalysts.[8] Generally, however, these materials lack the acid strength and stability of zeolites, and important commercial applications have not yet emerged.

ALPOs are synthesised using templates in a similar fashion to many of the zeolites. Typical templates for forming large pores are tetra-n-propyl ammonium ion and tri-n-propylamine. ALPO-5 has 12-ring windows with an aperture size of 0.8 nm (Figure 2.2).

More recently, extra large ALPOs have been prepared. The first of these was VPI-5 with 18-ring windows and an aperture size of 1.2–1.3 nm (Figure 2.2).[9]

Larger pore materials have followed, for example, cloverite, a 20 tetrahedral atom gallophosphate material that has a four leaf clover shaped pore opening

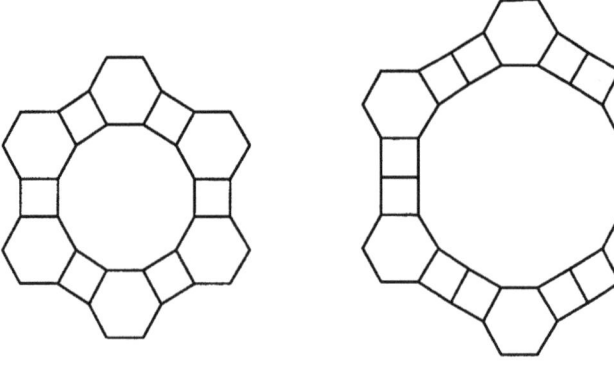

Structural (001) projection
of ALPO-5

Structural (001) projection
of VPI-5

Figure 2.2 *Mesoporous molecular sieves*

and a body diagonal dimension of about 3 nm.[10] The hope is that these large channelled structures will allow catalytic conversions to take place, involving molecules which are too large to enter the conventional zeolitic size channels.

6 Mesoporous Molecular Sieves

The relatively recent synthesis of a family of silica based molecular sieve materials (designated M41S) has attracted considerable interest because of the potential of these materials for use as larger pore catalysts. There have been several reports describing the synthesis of these materials and it is now recognised that there are a variety of routes by which they may be prepared.[11]

The acidic properties of mesoporous molecular sieves rely on the presence of active sites in their framework. In the case of MCM-41 active sites are generated by the introduction of heteroatoms into the structure. In particular, Brönsted acid sites are introduced by isomorphous substitution of Al for Si which is achieved by hydrothermal synthesis in which charged quaternary ammonium micelles are used as the template for charged alumino-silicate inorganic precursors.

The resulting (calcined) template free material (which contains charge-balancing inorganic cations) is ammonium exchanged and re-calcined to generate protons which give rise to Brönsted acid sites in the form of 'bridging' SiOHAl hydroxyl groups.

Tanev *et al.* have reported the synthesis of mesoporous materials *via* a route which involves self-assembly between neutral primary amines and neutral inorganic framework precursors.[12] The regularity of the pore structure in these materials has been illustrated by lattice images which show a honeycomb like structure. The system of channels of these molecular sieves produces solids with very high internal surface area and pore volume. This fact combined with the possibility of generating active sites within the channels produces a very unique type of acid catalyst. In the case of transition metal substituted M41S, the principal interest lies in their potential as oxidation catalysts, especially Ti and V substituted MCM and HMS type materials, and more recently synthesised large pore materials.[13]

7 Catalytic Applications of Zeolites and Related Materials

Zeolitic materials are most notably used in catalysis for shape-selective reactions. These reactions are mainly acid catalysed; however, base-catalysed and oxidation reactions have also been reported and will be discussed here.

Many of the reported examples of acid-catalysed reactions using zeolites are relevant to large-scale chemical manufacturing rather than fine or speciality chemical synthesis. These reactions will be briefly surveyed here since much of the chemistry may be more widely applicable, especially when applied to the larger pore zeotypes.

Alkylation of Aromatics

Electrophilic alkylation of aromatics can be carried out with a variety of alkylating agents such as alkenes, alcohols and halogenated hydrocarbons. Aromatic alkylation is a good example of a reaction where the shape selectivity of the zeolite plays an important role in controlling the distribution of products.

One of the most important industrial alkylations is the production of 1,4-xylene from toluene and methanol (Reaction 2). ZSM-5, in the proton exchanged form, is used as the catalyst because of its enhanced selectivity for *para* substituted products. *para*-Xylene is used in the manufacture of terephthalic acid, the starting material for the production of polyester fibres such as Terylene. The selectivity of the reaction over HZSM-5 occurs because of the difference in the rates of diffusion of the different isomers through the channels. This is confirmed by the observation that selectivity increases with increasing temperature, indicating the increasing importance of diffusion limitation. The diffusion rate of *para*-xylene is approximately 1000 times faster than that of the *meta* and *ortho* isomers.[14]

$$CH_3OH \;+\; \underset{}{\text{(toluene)}} \;\longrightarrow\; \underset{}{\text{(para-xylene)}} \;+\; H_2O \quad (2)$$

The HZSM-5 can be made even more selective towards *para*-xylene production by impregnation with an aqueous solution of orthophosphoric acid. Useful forms contain about 8.5% phosphorus by weight. Up to 97% of selective conversion to *para*-xylene has been achieved by this type of treatment. The spectacular selectivity for *para*-xylene formation is believed to originate from aluminium phosphates occupying sites in the pore openings of HZSM-5. This restricts the dimension of the pores slightly in comparison with HZSM-5 and makes it impossible for the *meta*- and *ortho*-xylenes to leave and only *para*-xylene is small enough to exit the zeolite structure. The P-modified H-ZSM-5 has been shown to contain fewer Brönsted acid sites than HZSM-5.[15] The Brönsted acidity can be restored completely by elution of the orthophosphoric acid with hot water. Only after steaming of the P-modified H-ZSM5 at elevated temperatures is an irreversible decrease of Brönsted acidity caused by de-alumination observed.[15]

The selective methylation of *meta*-xylene to produce 1,2,4-trimethylbenzene (TMB) has been studied by Raj *et al.*[16] The most effective catalysts were those based on the medium pore 10 ring MEL structure. They found that isomorphous substitution of framework Al for Ga or Fe significantly enhanced the yield of 1,2,4-TMB. The reason for the higher yields was attributed to the weaker acid sites on the Ga and Fe substituted materials compared to the Al analogs which allows the alkylation of xylene to compete more effectively with other reactions such as isomerisation of *meta*-xylene and conversion of methanol to aliphatics. The total yield of TMBs in the product could be increased significantly by methylating the equilibrium xylene mixture (instead of individual isomers) due to the suppression of isomerisation reaction.

One of the unique features of zeolites in alkylation reactions is their shape selectivity. In many zeolite-catalysed reactions, however, shape-selective catalysis occurring on the inside of the zeolite can be affected by non-selective catalysis on the external surfaces. Paparatto *et al.* have reported that during aromatic alkylation, the *para* isomer is formed selectively within the zeolite, whereas isomerisation occurred only on the external surfaces, decreasing *para* product selectivity.[17]

Several methods have been reported to deactivate the catalytic activity of the external surfaces of zeolites. Bhat *et al.* have modified the catalytic behaviour of ZSM-5 by chemical vapour deposition (CVD) of tetraethyl orthosilicate (TEOS).[18] The CVD technique does not affect the channel size or acidity of the zeolite but deactivates the external surfaces by coating them with an inert layer of silica. As a result, the shape selectivity of the zeolite is greatly enhanced.

Catalytic Cracking

While petrochemical processes are strictly beyond the scope of this book, the use of zeolites in cracking reactions will be briefly considered here since it reveals interesting results from studies on modifying zeolites.

Catalysts containing faujasite (FAU) are widely used for petroleum refining. The selectivity and activity of these catalysts in cracking is controlled by the number of aluminium acid sites per unit cell. Several studies have shown that the removal of exchanged sodium increases the catalytic cracking activity as would be expected for a solid Brönsted acid.[19] While sodium is a very strong poison for faujasite cracking catalysts, potassium was shown to be an even stronger poison. The deactivating effect of individual cations appears to be a function of their size and follows the order $K^+ > Na^+ > Li^+ > Mg^{2+} > H^+$ (least deactivating).

Direct fluorination of zeolites has been reported as a method for increasing their acidity for hydrocarbon cracking reactions. Lok *et al.* treated various zeolites with fluorine gas and reported zeolite dealumination and stabilisation and in certain cases an increase in n-butane cracking activity.[20] It appears that the optimum fluorine content for maximum cracking activity is about 1% m/m.

The expected result of dealumination is a decrease in the total number of Brönsted acid sites, whereas the number of strong Brönsted acid sites increases relative to the number of aluminium atoms because of an increase in the number of isolated Al. The number of Lewis acid sites also increases because of an increase in the non-framework aluminium content.[21]

For catalytic cracking, not all zeolites are used in the decationised or proton exchanged form; it is also quite common to replace the original Na$^+$ ions with lanthanide ions such as La^{3+} or Ce^{3+}. It is interesting to note that the first commercial zeolite for cracking was a rare earth substituted form of zeolite X.[22]

The high catalytic activity of lanthanum-exchanged zeolites has been attributed to the presence of polyvalent lanthanum ions in the form of $[La(OH)_2La]^{4+}$ or $La(OH)_2^+$ species in the large zeolite cavities, withdrawing electrons from the framework OH groups making the protons more acidic.[23]

Fischer–Tropsch Synthesis

There is at present much interest in the use of solid ruthenium catalysts for Fischer–Tropsch synthesis.[24] It has been found that the maximum chain growth in the synthesis reaction is strongly affected by the size of the ruthenium particles. The smaller the particles, the lower the molecular weight of the products. Specifically it has been found that the maximum petroleum production should result if the crystallite size can be controlled in the range of 3–4 nm.

The usual method for preparing ruthenium supported on zeolites is by cationic exchange with $Ru(H_2O)_6^{2+}$ and $Ru(NH_3)_6^{2+}$. Highly dispersed ruthenium within the zeolite structure was obtained even when the sample was heated in an inert atmosphere due to auto-reduction by the ammonia released by decomposition of the complex. This would appear to be a fairly general phenomenon when using amine complexes. Further heating in hydrogen results in crystalline growth particularly in the presence of water. Oxidation of the highly dispersed ruthenium above 700 °C led to the growth of very large crystallites on the external surfaces of the zeolite.

Aromatisation

Catalytic aromatisation of aliphatic hydrocarbons was first described by researchers at Mobil.[25] In this process, termed 'M-2 forming', alkanes from ethane to high boiling point naphthalenes can be aromatised. The most effective catalyst for aromatisation was found to be the medium pore HZSM-5.[26] Large pore zeolite and amorphous silica–alumina with broad pore size distributions gave only low yields of aromatics due to rapid coke formation.

Using the catalytic conversion of propane to aromatics as a model reaction system Kwak *et al.* have studied the effect of adding Ga and Pt promotors to HZSM-5.[27] The intrinsic dehydrogenation activity at low conversions increases in the order Ga < PtGa < Pt. At higher conversions the reverse order is found for the production of aromatics. In spite of its intrinsically high dehydrogenation activity, Pt was found not to be a suitable promotor of HZSM-5 in aromatisation reactions because of rapid deactivation due to coke build-up. The Ga-containing and Ga–Pt zeolites were much more resistant to deactiviation. The authors suggested that the added metals (Ga, Pt) may play an independent, additive role as propane dehydrogenation catalysts in addition to the strong acid sites of HZSM-5, the combination acting as a classical bifunctional catalyst. This view does not rule out the possibility that the added Ga may replace some of the zeolite protons, or that the dehydrogenation

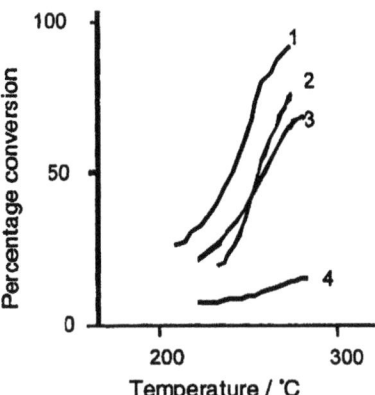

Figure 2.3 *Dehydration of alcohols using zeolites: dehydration of* (1) *butan-2-ol on zeolite-X;* (2) *n-butan-1-ol on zeolite-X;* (3) *butan-1-ol on zeolite-A;* (4) *butan-2-ol on zeolite-A*

activity of the Ga may be increased due to greater dispersion within the zeolite cavity.

Alcohol Dehydration

The dehydration of alcohols to alkenes over zeolite-A provides an important example of the reactant selectivity of zeolites. Under conventional conditions butan-2-ol forms the more stable carbonium ion and therefore dehydrates much more easily than butan-1-ol.[28] Using zeolite-A, however, only butan-1-ol is small enough to enter the zeolite and access the active acid sites and undergo dehydration. The butan-2-ol is excluded and is thus not converted. Zeolite-X has windows large enough to admit both alcohols and both undergo conversion to the corresponding alkene. These results are summarised in Figure 2.3. It is interesting to note that at higher temperatures curve (4) begins to rise. This is because the lattice vibrations increase with temperature, making the pore opening in zeolite-A slightly larger and thus beginning to admit butan-2-ol. The low conversion at moderate temperatures is thought to result from reactions taking place on external sites.

Methanol Synthesis

Methanol synthesis from 'syngas' (CO and H_2) is another example of a large-scale process that can be catalysed at high temperature using zeolites, as has been shown to take place on Pd–SAPO.[29] This is a useful process since the methanol can then be converted to petrol using ZSM-5 zeolites in the MTG process developed by Mobil. The distribution of hydrocarbon products is largely determined by the pore size of the SAPO. Pd/SAPO-5 catalysed significant C2 hydrocarbon formation whilst high yields of C2–C4 alkanes were observed with Pd/SAPO-34.

Base Catalysis

It has been reported that zeolites can be used as base catalysts when exchanged with alkali metal ions. The base strength is inversely related to the charge to radius ratio of the compensating cation, *i.e.* the larger cation the stronger the basicity of the associated framework oxygen of the zeolite.

The Cs$^+$-exchanged zeolites, which are the most basic, have been shown to catalyse the Claisen Schmidt condensation between substituted 2-hydroxy-acetophenones and substituted benzaldehyde to give the 2'-hydroxychalcone structure (Reaction 3).[30]

Chalcones are important intermediates in the synthesis of flavanoids and are used industrially in bactericides, antibiotic drugs and UV-stabilisers in plastics. Other base catalysts such as magnesium *t*-butoxide and barium hydroxides have been used to perform the synthesis.[31] However, the Cs$^+$-exchanged zeolites offer a more environmentally friendly alternative route.

Cs$^+$- and Na$^+$-exchanged MCM-41 type materials also have basic character and have been found to be active towards the base catalysed Knoevenagel condensation of benzaldehyde with ethyl cyanoacetate (Reaction 4).[32] The Cs$^+$- and Na$^+$-exchanged samples were prepared by repeated exchange of the hydrogen form of MCM-41 with an aqueous solution of appropriate chloride salt (0.5 mol dm^{-3}) at room temperature. The Cs$^+$-exchanged sample was considerably more basic and therefore more active than the Na$^+$-exchanged sample.

Oxidation

Alkene oxidation over transition metal exchanged zeolites has been of recent interest. Yu and Kevan have studied the partial oxidation of propene to acrolein over Cu^{2+} and Cu^{2+}/alkali–alkaline earth exchanged zeolites.[33] In both systems cuprous ions are believed to be the catalytically active sites for oxidation. The degree of reoxidation and rehydration of these cations was found to be important in regenerating the active sites.

Titanium silicate (TS-1) which has a structure similar to the zeolite ZSM-5 has been shown to catalyse a number of synthetically important oxidations with hydrogen peroxide under mild conditions.[34] A useful feature of the TS-1 catalyst is its enhanced product selectivity in oxidation reactions, for example, cyclohexane is selectively oxidised to cyclohexanone inside the pores of TS-1. On the external surfaces where there is little steric control cyclohexane is oxidised to the dicarboxylic acid. Spinace and co-workers have shown that these external reactions can be prevented by the addition of an antioxidant such as 2,6-di-*tert*-butyl-4-methylphenol (BHT) but which does not interfere with the internal reactions since it is too bulky to enter the pores of the TS-1.[35]

Substituted mesoporous silicas are very promising catalysts for the oxidation of arylamines in the liquid phase. Indeed Gontier and Tuel have reported that the performance of TS-1 was considerably poorer than large pore zeolite Ti- and V-substituted molecular sieves for the oxidation of aniline.[36] At low oxidant/aniline ratios it was found that azoxybenzene was the major product using Ti-substituted molecular sieves. In contrast, V-substituted molecular sieves were very selective towards the conversion of aniline to nitrobenzene. The difference between the Ti and V molecular sieves was attributed to the greater number of active oxidising sites in the V-HMS, leading to further oxidation of azoxybenzene into nitrobenzene.

Rearrangements

Titanium silicate (TS-1) has also been used to catalyse the Beckmann rearrangement of cyclodehexanone oxime to ε-caprolactam.[37] The caprolactam is an important starting material for the manufacture of nylon fibres. The normal industrial method is to use sulfuric acid as the catalyst. However, the use of sulfuric acid for this reaction does have its disadvantages, such as the formation of low value by-products such as ammonium sulfate, reactor corrosion and environmental hazards associated with its disposal. Using TS-1 alleviates these problems and gives over 90% yield of ε-caprolactam. Small amounts of high boiling condensation products were also produced but these may be easily separated from the ε-caprolactam by fractional distillation. In comparison with other catalysts sharing the MFI structure, such as silicalite, the TS-1 gave higher conversions of oxime and better selectivity towards ε-caprolactam. The yield of ε-caprolactam was found to be dependent on the Ti content of the TS-1.

Ammoxidation

Ammoxidation of cyclic ketones over titanium silicates TS-1 has been performed.[38] The reactivity of the cyclohexanones and methylcylcohexanones over TS-1 followed the order cyclohexanone > 2-methylcyclohexanone = 3-methylcyclohexanone > 2,6-dimethylcyclohexanone, reflecting the difference in the diffusion rates of the products inside TS-1. From the relative reactivity of dimethylcyclohexanone isomers it was suggested that the steric hindrance of the substituent methyl group to the access of carbonyl group inside the catalyst decreases in the order b-equitorial > a-equitorial > b-axial > a-axial.

Epoxidation

Hydrogen peroxide and organic hydroperoxides are relatively poor oxidants in the absence of radical initiators or other specific reagents. No reaction occurs with alkenes unless a reagent capable of producing electrophilic intermediates, such as a peracid or metal peroxo complex, is used.

The major breakthrough was the discovery that titanium silicate could used as an efficient epoxidation catalyst.[39] The reaction with TS-1 may be performed under mild conditions, for example at room temperature in dilute aqueous or methanolic solutions.

The most widely accepted mechanism for TS-1 catalysed epoxidation is the peracid-like mechanism in which the active epoxidising species acts as the electrophile. In addition to the mild conditions TS-1 offers the advantage of shape selectivity which results from the active sites being situated in a pore system of approximately 0.55 nm in diameter. The branched and cyclic alkenes react much more slowly than the linear alkenes

8   Future Trends in Zeolite Catalysts

Further catalytic uses of zeolites and related materials include polymerisation of alkenes as well as the development of basic zeolitic materials generated by the incorporation of alkali metal ions. Gallium- and boron-substituted zeolites have already been shown to be useful catalysts in wide variety of reactions (Table 2.2) and undoubtably these will be followed by novel zeotypes including mesoporous materials with other catalytically active elements within their frameworks.[40]

9   New Developments in the Context of Clean Synthesis

It is still true to say that zeolites have found relatively little use in the liquid phase synthesis of organic compounds due to their small pore size and related

Table 2.2 *Reactions catalysed by boron-
and gallium-substituted ZSM-5*

Catalyst	Type of reaction
BZSM-5	disproportionation
BZSM-5	isomerisation
BZSM-5	hydroisomerisation
BZSM-5	cracking
GaZSM-5	aromatisation

diffusional limitations. However in some cases they can offer unique advantages in term of excellent selectivity towards, for example, mono-substitution and positional isomerism. Exceptionally high selectivity is a requirement in the synthesis of precursors to many bioactive compounds and is becoming an ever-increasingly important property of any new manufacturing process as waste becomes less and less acceptable. A significant proportion of the new developments in the context of the use of zeolites in organic synthesis has been driven by the goals of clean synthesis. Some examples illustrating this trend are given below.

Owing to their numerous applications as fine chemicals for the synthesis of bioactive compounds such as pesticides and pharmaceuticals, isomerically pure chloroaromatics are very valuable materials. *t*-Butyl hypochlorite/HNa faujasite in acetonitrile represents an efficient and highly regioselective system of mono-chlorination of a wide range of mono- and disubstituted aromatic substrates in mild conditions (Reaction 5).

$$\text{(5)}$$

Partially protonated faujasite X is far superior to amorphous silicas and to other zeolites in terms of efficiency and regioselectivity (Table 2.3). Advantages of *t*-butyl hypochlorite over other chlorinating reagents have been demonstrated.[41] The methodology is applicable to a range of substituted benzenes with

Table 2.3 *Influence of the catalyst on the chlorination of toluene*

Catalyst	Conversion of toluene (%)	Products
HCaA	0	
NaZSM-5	7	52% *p*-chlorotoluene
		48% dichlorotoluenes
HNaZSM-5	20	47% *p*-chlorotoluene
		53% *o*-chlorotoluene
HNaMordenite	15	40% *p*-chlorotoluene
		60% *o*-chlorotoluene
KL	0	
NaX	11	65% *p*-chlorotoluene
		35% *o*-chlorotoluene
HNaX	90	65% *p*-chlorotoluene
		35% *o*-chlorotoluene
Kieselgel 60 (Silica)	80	40% *p*-chlorotoluene
		60% *o*-chlorotoluene

Table 2.4 *Chlorination of substituted benzenes with t-butyl hypochlorite/HNa-X zeolite in acetonitrile*

Substituent	Product yield (%)	para/ortho *chlorination ratio*
OMe	100	82:18
Me	100	82:18
Et	100	90:10
i-Pr	90	80:20
t-Bu	99	98:2
Ph	86	86:14
Cl	95	97:3
Br	75	97.3

particularly high selectivities towards *para*-chlorination with chlorobenzene, bromobenzene and *t*-butylbenzene (Table 2.4).

The Beckmann rearrangement has been extensively studied for many years. The most important industrial examples are the conversion of cyclohexanone and cyclodecanone oximes into the corresponding lactams, which are the precursors for the fabrication of nylon 6 and nylon 12 respectively. The classical large-scale rearrangement of cyclohexanone oxime is carried out in the liquid phase using concentrated sulfuric acid. Although it is a simple and quite convenient method employing a very common and inexpensive acid, the use of large quantities of hazardous fuming sulfuric acid, the associated corrosion problems and, particularly in the context of this book, the large volumes of waste resulting from the neutralisation of the used acid (to ammonium sulfate) make it environmentally questionable. A large number of heterogeneous catalysts have been tested in these reactions. Typically they have been run under vapour phase conditions, at temperatures of 250–350 °C. Such high temperatures can lead to a decrease in reaction selectivity and rapid catalyst deactivation. Using environmentally friendly solid acid catalysts in a liquid phase reaction is an important goal. Progress has been made in this direction using beta-zeolites.[42] Reaction occurs at 130 °C, although when the zeolite does not contain framework Al and internal silanols, no appreciable conversion was observed. When the catalyst has internal silanols but no framework Al, oxime conversion is obtained but the selectivity to the corresponding amide can be low. In the case of beta-zeolite without silanol groups but with framework Al, conversion and selectivity are high. During the Beckmann rearrangements studied, both the amide and the parent ketone can be observed as products (Reaction 6).

(6)

The results for different zeolites are shown in Table 2.5.

Table 2.5 *Beckmann rearrangement of cyclohexanone oxime over zeolites*

Zeolite	Conversion	Proportion of product
Aerosil	5	98% ketone
H-ZSM5 (high Al, high silanol)	67	95% amide
Amorphous silica	0.1	72% amide
Beta-D (low Al, high silanol)	38	98% amide
Beta-ND1 (low Al)	0	
H-beta-ND (high Al, low silanol)	54	98% amide
H-beta-D (high Al, high silanol)	68	98% amide

The adage that the best protecting group is no protecting group is very true but the reality is that there is still a need to protect reactive functional groups as a method of improving selectivity. It is therefore essential for clean synthesis that the protection and deprotection steps generate very little waste. Zeolites can be used for the efficient formation of dithianes as protecting groups for carbonyl compounds for carbonyl compounds.[43] The methodology can be extended to the synthesis of phenylhydrazones and 2,4-dinitrophenylhydrazones (Figure 2.4). When the non-acidic zeolite NaY is employed as the catalyst, no derivative of the carbonyl compound is obtained. This rules out any contribution of the basic zeolite framework and indicates the need for acidic sites. Indeed, when acidic zeolites are used, such as HY, CaY and MgY, the formation of the derivatives is smooth and clean and the product yields are excellent (Table 2.6).

Friedel–Crafts reactions continue to represent one of the greatest challenges for clean synthesis. They are widely used to manufacture an enormous range of important chemical products and intermediates, ketones being among the most important with applications in pharmaceutical and agrochemical products, flavours and fragrances. Traditional methods of manufacture are based on

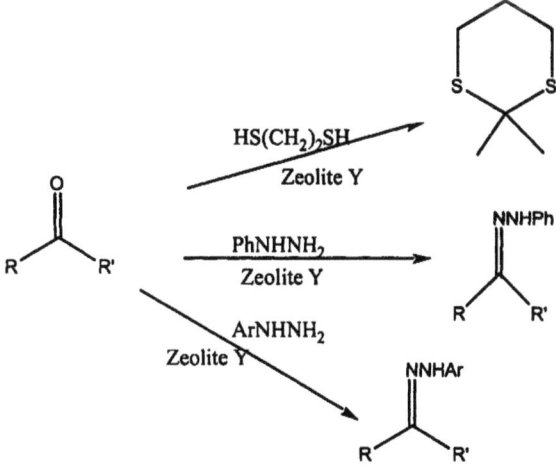

Figure 2.4 *Zeolite-catalysed protection of carbonyl compounds*

Table 2.6 *Zeolite Y-catalysed protection of carbonyl compounds*

| Carbonyl compound | Zeolite | Derivative yield (%) | | |
		Phenylhydrazone	*Dinitrophenylhydrazone*	*Dithiane*
Benzaldehyde	HY	90	92	96
	CaY	87	92	91
Acetophenone	HY	93	87	92
	MgY	80	82	83
Cyclohexanone	HY	87	89	87
	CaY	83	85	89

strong Lewis and Brönsted acids, including most famously aluminium chloride. Apart from health and safety problems in handling such hazardous reagents, the volume and toxicity of the waste generated in the work-up stages of such reactions is unacceptable. The use of benign and recoverable solid acids would greatly alleviate these problems and added reaction selectivity, typically through the use of porous solids, will be of further advantage. In this context, the zeolite HBEA has been shown to have good catalytic activity at least for the Friedel–Crafts reactions of activated substrates such as anisole and *m*-xylene.[44,45] The effect is enhanced by the presence of extraframework aluminium.[46] Acid treatment increases the catalytic activity of the outer surface of the zeolite due to the extraction of catalytically active extraframework alumina species out of the micropores and due to the formation of silanol species. The effects on the acetylation of 2-methoxynaphthalene are shown in Figure 2.5.

The Fries rearrangement provides another important route to ketones. For example, the rearrangement of phenyl acetate to hydroxyacetophenones can be used on route to *p*-hydroxyacetanilide, which is an important painkiller (paracetamol). The use of solid catalyst to effect this reaction has had limited success however, typically due to low selectivity and easy catalyst deactivation.

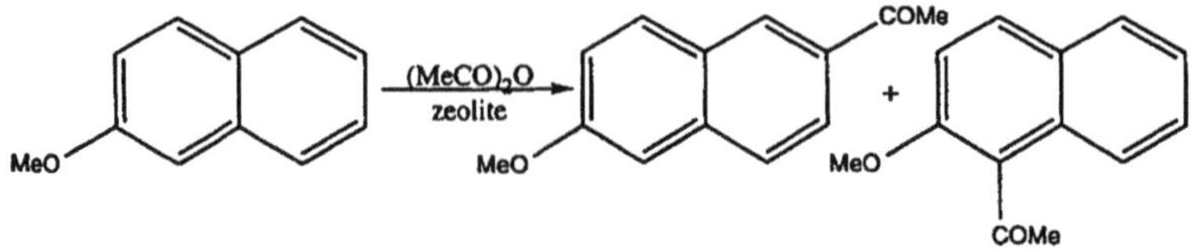

Zeolite	Conversion (%)	Selectivity
HBEA (untreated)	30	19:81
HBEA (calcined)	39	53:47
HBEA (acid-treated)	54	10:90

Figure 2.5 *Zeolite-catalysed acylation of 2-methoxynaphthalene*

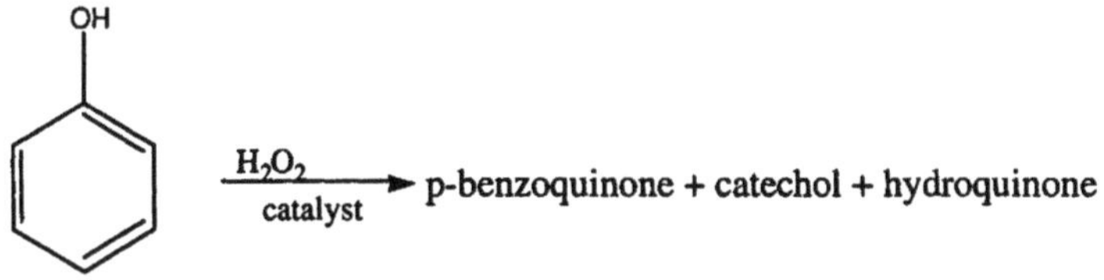

Figure 2.6 *Hydroxylation of phenol catalysed by Sn-mesoporous materials*

Catalyst	Conversion	Product distribution
Sn-MCM-41	23	24:58:18
Sn-impregnated MCM-41	1	50:25:25
SnO$_2$	0.5	54:24:22

At 150 °C with zeolites such as the ultrastable faujasite H-USY, conversions up to about 40% and selectivities of up to about 80% can be achieved.[47] The best selectivities are observed in phenol as solvent. The main by-product is phenol, which may be formed *via* the decomposition of the starting material or the deacetylation of the product. This inevitably seems to result in catalyst deactivation.

The advantages of mesoporous materials for acid catalysed reactions are based on the presence of large regular pores, which allow the diffusion of reactants, and conversely the diffusion of the products out, minimising side-reactions and catalyst deactivation through poisoning. These characteristics are very important in the clean synthesis of fine chemicals, which involve bulky reactants or products, and where the liquid phase environment commonly employed maximises diffusional problems. Their use in chemicals synthesis is, however, still in its infancy. One area where their value as catalysts has been demonstrated is in the synthesis of long chain alkyl glucosides which have excellent surfactant properties, biodegradability and low degree of oral and skin toxicity. They are used as food emulsifiers, pharmaceutical dispersing agents and cosmetic surfactants. These compounds can be synthesised from glucose and butan-1-ol on Al-MCM-41 mesoporous solid acids.[48] Interestingly, a higher concentration of acid sites does not guarantee a better catalytic performance and the adsorption–desorption properties of the substrates and products play a very important role in the reaction when these molecules have very different polarities. Pore size is important. The larger the pore at the same Al-level, the more active is the catalyst. The catalysts deactivate on use, however.

As with zeolites, other metals can be incorporated into the structure of mesoporous solids to give a wider spectrum of applications. A good example of this is the tin-containing MCM analogues, stannosilicates that can catalyse various oxidation reactions under mild conditions.[49] Whereas the tin-free analogues are inactive as catalysts in the hydroxylation of phenol and 1-naphthol, the tin materials give reasonable turnover numbers in the reactions (Figures 2.6 and 2.7). Interestingly, the impregnated versions showed poor

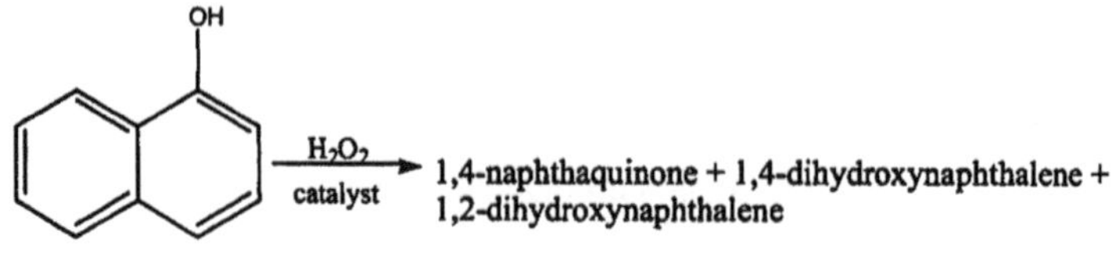

Catalyst	Conversion	Product distribution
Sn-MCM-41	19	82:13:5
Sn-impregnated MCM-41	1	50:50:0
SnO_2	0.5	60:20:20

Figure 2.7 *Hydroxylation of naphthalene using Sn-mesoporous materials*

Table 2.7 *Selective Diels–Alder reactions using solid catalysts*

Catalyst	*Yield (%)*	*exo:endo*
None	6	79:21
BF_3-Et_2O	39	96:4
Zn^{2+}-K10	46	94:6
Zn^{2+}-K10	46	94:6
Zn^{2+}-Al-MCM-41	90	95:5

activity presumably because of the need for highly active Sn^{4+} centres, which are incorporated into the framework of the Sn-MCM as opposed to the impregnated materials. The Sn-MCM-41 materials are also active in the epoxidation of norbornene using *t*-butyl hydroperoxide (TBHP). Reaction reaches a maximum in terms of substrate conversion after about 10 h (*ca.* 28% conversion) presumably due to exhaustion of the hydroperoxide. At this point the selectivity to the *exo*-2,3-epoxynorbornene is about 85%. Once again the Sn-impregnated sample is much less active (2% conversion).

The Diels–Alder reaction has been shown to be subject to catalysis by a wide range of solid catalysts (see Chapter 4 for some examples). Acidic mesoporous aluminosilicates can be used to catalyse selective Diels–Alder reactions such as that between cyclopentadiene with methyl acrylate. The zinc-exchanged version of the material is particularly effective and compares well to other more established solid acids such as the ion-exchanged clay Zn^{2+}-K10 as well as homogeneous catalysts such as boron trifluoride (Table 2.7).[50]

References

1 D.W. Breck 'Zeolite Molecular Sieves', John Wiley, New York, 1974.
2 A. Dyer 'An Introduction to Zeolite Molecular Sieves', John Wiley, New York, 1988.
3 J.M. Newsam, *Science*, 1986, **231**, 1093.
4 D.A. Whan, *Chem. Br.*, 1981, 532.

5 R.M. Barrer 'Hydrothermal Chemistry of Zeolites', Academic Press, New York, 1982.
6 G.T. Kerr, *Sci. Amer.*, July, 82, 1989.
7 J.M. Thomas, *Sci. Amer.*, April, 82, 1992.
8 G.A. Ozin, A. Kuperman and A. Stein, *Angew. Chem., Int. Ed. Engl.*, 1989, **28**, 359.
9 M.E. Davis, C. Saldarriaga, C. Montes, J. Garces and C. Crowder, *Nature*, 1988, **331**, 698.
10 M. Estermann, L.B. McClusker, C. Baerlocher, A. Merroche and H. Kessler, *Nature*, 1991, **52**, 320.
11 C.T. Kresge, M.E. Leonowicz, W.J. Roth, J.C. Vartuli and J.S. Veck, *Nature*, 1992, **359**, 710.
12 P.T. Tanev, M. Chibwe and T.J. Pinnavaia, *Nature*, 1994, **368**, 321.
13 A. Corma, M.T. Navarro and J. Perez-Pariente, *J. Chem. Soc., Chem. Commun.*, 1994, 147.
14 R. Kumar, G.N. Rao and P. Ratnasamy, *Stud. Surf. Sci. Catal.*, 1989, **49**, 1141.
15 G. Lischke, R. Eckelt, H.-G. Jerschkewitz, B. Parlitz, E. Schreier, W. Storek, B. Zibrowius and G. Öhlmann, *J. Catal.*, 1991, **132**, 229.
16 A. Raj, S. Reddy and R. Kumar, *J. Catal.*, 1992, **138**, 518.
17 G. Paparatto, E. Morettic, G. Leofanti and F. Gatti, *J. Catal.*, 1987, **105**, 227.
18 Y.S. Bhat, J. Das, K.V. Rao and A.B. Halgeri, *J. Catal.*, 1996, **159**, 368.
19 R. Kumar, W.-C. Cheng, K. Rajagopalan, A.W. Peters and P. Basu, *J. Catal.*, 1994, **143**, 594.
20 B.M. Lok, F.P. Gorstema, C.A. Messina, H. Rastelli and T.P. Izod, *ACS Symp. Ser.*, 1983, **218**, 41.
21 A.G. Panov, V. Gruver and J.J. Fripiat, *J. Catal.*, 1997, **168**, 321.
22 P. Marynen, A. Maes and A. Cremers, *Zeolites*, 1984, **4**, 287.
23 R. Carvajal, P. Chu and J. Lunsford, *J. Catal.*, 1990, **125**, 123.
24 Y. Kikuzuno, S. Kagami, S. Naito and K. Tamaru, *Chem. Lett.*, 1981, 1249.
25 N.Y. Chen and T.Y. Yan, *Ind. Eng. Chem. Process Des. Dev.*, 1986, **25**, 151.
26 P.B. Weisz, W.O. Haag and P.G. Rodewald, *Science*, 1979, **206**, 57.
27 B.S. Kwak, W.M.H. Sachtlet and W.O. Haag, *J. Catal.*, 1994, **149**, 465.
28 P.B. Venuto and P.S. Landis, *Adv. Catal.*, 1969, **18**, 259.
29 R. Thompson, C. Montes, M.E. Davis and E.E. Wolf, *J. Catal.*, 1990, **124**, 401.
30 M.J. Climent, A. Corma, S. Iborra and J. Primo, *J. Catal.*, 1995, **151**, 60.
31 J.L. Guthrie and N. Rabjohn, *J. Org. Chem.*, 1957, **22**, 1761957.
32 A. Corma, *Chem. Rev.*, 1997, **97**, 2373.
33 J. Yu and L. Kevan, *J. Phys. Chem.*, 1991, **95**, 6648.
34 R.A. Sheldon and J. Dakka, *J. Catal. Today*, 1994, **19**, 215.
35 E.V. Spinace, H. Pastore and U. Schuchardt, *J. Catal.*, 1995, **157**, 631.
36 S. Gontier and A. Tuel, *J. Catal.*, 1995, **157**, 124.
37 A. Thangaraj, S. Sivasanker and P. Ratnasamy, *J. Catal.*, 1992, **137**, 252.
38 T. Tatsumi and N. Jappar, *J. Catal.*, 1996, **161**, 570.
39 M.G. Clerici and P. Ingallina, *J. Catal.*, 1993, **140**, 71.
40 C.T.W. Chu, G.H. Kuehl, R.M. Lago and C.D. Chang, *J. Catal.*, 1985, **93**, 451.
41 K. Smith, M. Butters, W.E. Paget, D. Goubet, E. Fromentin and B. Nay, *Green Chemistry*, 1999, 83.
42 M.A. Camblor, A. Corma, H. Garcia, V. Semmer-Herledan and S. Valencia, *J. Catal.*, 1998, **177**, 267.
43 A. Lalitha, K. Pitchumani and C. Srinivasan, *Green Chemistry*, 1999, 173.
44 K. Smith, Z. Zhenhua, L. Delaude and P.K.G. Hodgson, in *Proceedings 4th International Symposium on Heterogeneous Catalysis Fine Chemicals*, Basel, 1996.
45 H.K. Heinichen, *Thesis*, RWTH-Aachen, 1999.
46 H.K. Heinichen and W.F. Hoelderich, *J. Catal.*, 1999, **185**, 408.
47 A. Heidekum, M.A. Hamer and W.F. Hoelderich, *J. Catal.*, 1998, **176**, 260.

48 M.J. Climent, A. Corma, S. Iborra, S. Miquel, J. Primo and F. Rey, *J. Catal.*, 1999, **183**, 76.
49 K. Choudhari, T.K. Das, P.R. Rajmohanan, K. Lazar, S. Sivasanker and A.J. Chandwadkar, *J. Catal.*, 1999, **183**, 281.
50 M. Onaka and R. Yamasaki, *Chem. Lett.*, 1998, 259.

Clay Materials

1 Introduction

Most applications for clays as catalysts have been in acid catalysed reactions[1,2] but as an understanding of clays has developed, they have been used as the basis for a large number of more varied catalysts.[3,4]

2 Structure of Clays

Clays are naturally occurring layered silicates. They are essentially crystalline materials of very fine particle size, nominally < 2 μm in diameter. The crystal structure of clay minerals was first proposed by Pauling[5] and the validity of his model was confirmed by the powder X-ray diffraction technique.[6] According to Pauling's model the structure can be described as shown in Figure 3.1. SiO_x tetrahedra are linked in two dimensions to form repeated hexagonal patterns of basal oxygens, *i.e.* tetrahedral sheets. $AlO_4(OH)_2$ in octahedra are also linked to form octahedral sheets.

The octahedral sheet is further divided into two types: the dioctahedral type

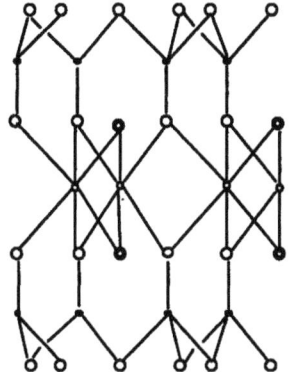

∘   aluminium

⊙   hydroxyl

•   silicon

○   oxygen

Figure 3.1   *Structure of a 2 : 1 type clay*

where two thirds of octahedral cation sites are occupied by trivalent cations (Al^{3+}, Fe^{3+}, *etc.*) and the trioctahedral type where all the sites are fully occupied by divalent cations (Mg^{2+}, *etc.*); when one octahedral sheet and two tetrahedral sheets are joined together the 2:1 triple layer is formed.[6]

The silicate layer is an electrically neutral structure. If part of tetrahedral Si^{4+} or octahedral Al^{3+} is isomorphously substituted by lower valency cations the sheet becomes negative. In this case the clay will have exchangeable cations between every layer to compensate for the negative charge of the layers. The cation exchange capacity (CEC) of a clay is equivalent to the layer charge and is dependent on the degree of isomorphous substitution.

Table 3.1 summarises the chemical compositions of the 2:1 type clays. The primary classification is based on layer charge and has four groups: pyrophyllite, smectites, vermiculites and micas. Pyrophyllite structures are electrically neutral and show no cation exchange capacity. Smectites have CECs between 60 and 100 millimole equivalents $100 \ g^{-1}$. Vermiculites have CECs in the range of 100–150 millimole equivalents $100 \ g^{-1}$. Micas have a large theoretical CEC of approximately 250 millimole equivalents $100 \ g^{-1}$ but natural micas do not swell in water and the interlayer K^{+} ions are firmly trapped between layers in a non-exchangeable form. Each group is divided according to whether the octahedral sheet is dioctahedral or trioctahedral.

Montmorillonite, which is the most commonly used clay for catalysing organic reactions, is a smectite clay in which the unit layer is composed of one octahedral sheet containing aluminium and magnesium ions sandwiched between two sheets of silica oxygen tetrahedra. In between the layers are hydrated metal ions, usually Na^{+} or Ca^{2+} in the natural materials.

The layers extend in the *a* and *b* axis directions and are stacked one above the other in the *c* direction. The thickness of one layer is about 0.96 nm but the

Table 3.1 *Chemical compositions of the 2:1 type clays*

Mineral group	Subgroup	
	Dioctahedral	*Trioctahedral*
Pyrophyllite	Pyrophyllite $[Al_2](Si_4)O_{10}(OH)_2$	Talc $[Mg_3](Si_4)O_{10}(OH)_2$
Smectite $(0.25 < x < 0.6)$	Montmorillonite $M_x[Al_{2-x}Mg_x](Si_4)O_{10}(OH)_2$ Beidellite $M_x[Al_2](Si_{4-x}Al_x)O_{10}(OH)_2$	Hectorite $M_x[Mg_{3-x}Li_x](Si_4)O_{10}(OH)_2$ Saponite $M_x[Mg_3](Si_{4-x}Al_x)O_{10}(OH)_2$
Vermiculite $(0.6 < x < 0.9)$	Dioctahedral vermiculite $M_x[Al_2](Si_{4-x}Al_x)O_{10}(OH)_2$	Trioctahedral vermiculite $M_x[Mg_3](Si_{4-x}Al_x)O_{10}(OH)_2$
Mica	Muscovite $K[Al_2](Si_3Al)O_{10}(OH)_2$	Phlogopite $K[Mg_3](Si_3Al)O_{10}(OH)_2$

interlayer spacing can vary according to the extent of hydration.[7] Powder X-ray diffraction can be used to measure the layer spacing.

3 Methods of Increasing the Catalytic Activity of Clays

Smectite clays have three important properties related to catalytic activity, intercalation, swelling and cation exchange capacity. When smectites are immersed in water, both intercalation of water molecules and swelling occur. The suspended clay can also freely exchange its interlayer cations for other cations in solution. Interlayer water molecules are dissociated producing protons and exhibit Brönsted acidity (Reaction 1).[8]

$$[M(H_2O)_6]^{n+} \rightarrow [MOH(H_2O)_5]^{(n-1)+} + H^+ \qquad (1)$$

The higher the electronegativity of the interlayer metal cation M^{n+} the stronger the acid sites. When montmorillonite is ion-exchanged with Al^{3+} ions, it displays a very high activity towards several Brönsted acid catalysed reactions, comparable to that of 98% H_2SO_4.[9]

The intrinsic acid character of clay materials has long been applied to various acid catalysed reactions. Industrially acid-activated clay (acid treated montmorillonite) was used as a cracking catalyst to obtain gasoline in the 1930s and the catalyst was in practical use until the mid-1960s.[10] The most serious disadvantage of such clay minerals in this application is the collapse of the interlayer space due to the elimination of water of hydration when the catalyst is heated above approximately 200 °C.[11] Under these circumstances only the external surfaces of the clay with only a small surface area (10–20 $m^2 g^{-1}$) is available for catalytic cracking. This disadvantage and the advent of more thermally stable synthetic aluminosilicates led to the replacement of clays as cracking catalysts.

4 Clay-supported Metal Catalysts

The high surface area of clays also makes them particularly attractive as catalyst and reagent supports. Discussion in this chapter will be largely confined to clay-based catalysts in which metal complexes or other ions are specifically incorporated in the clay matrix. Other clay-based catalytic materials will be discussed in Chapter 4.

Metal catalysts are frequently used in a supported rather than pure form. The role of the support is to maximise the surface area of the active phase by providing a large area over which it may be dispersed and hence in a highly active form when expressed as a function of the weight of the active component. This feature of supported catalysts is especially important with regard to precious metal catalysts because it allows more effective utilisation of the metal than can be achieved in the bulk metal systems. The high dispersion also helps to prevent the agglomeration and sintering of small metal particles under reaction conditions and allows the active phases to be moulded into coarse particles suitable for fixed reactor beds. Typical substances which find use as high surface area supports are alumina, silica and activated carbon, all of which have specific surface areas between 200 and 1000 m^2 g^{-1}. For clays to be useful as catalyst supports they must exhibit similar surface areas. Only the smectite clays can approach these values and then only if the internal surfaces are accessible in addition to the external surface area of the clay platelets.

The smectite clays do, however, have some important features which make them particularly attractive as catalyst supports. In addition to their high intrinsic surface area, their laminar structure may confer size and shape selectivity to the resultant catalysts. Another important feature is the negative charge on the silicate layers which may be able to polarise reactant molecules and enhance catalytic activity. Finally the intrinsic acidity of clay minerals provides the catalyst with bifunctionality. This may be useful for example in stabilising intermediate carbocations which would otherwise deprotonate.

A wide range of techniques may be employed for the incorporation of a catalytically active component into clay supports. An outline of the two most important techniques is given below as an introduction to later sections in this chapter, which describe the more important chemical and physical factors involved in the dispersion of metal salts onto clays and their influence on the activity and selectivity of the catalyst system. Methods for supporting species onto high surface area materials are described in some detail in Chapter 4.

Impregnation—Impregnation as a means of supported catalyst preparation is achieved by filling the pores of a support with a solution of the metal salt from which the solvent is subsequently evaporated. The catalyst is prepared either by spraying the support with a solution of the metal compound or by adding the support material to a solution of a suitable metal salt, so that the required amount of active component is incorporated into the support without the use of excess solution. This is then followed by drying and subsequent decomposition of the metal salt at an elevated temperature, either by thermal decomposition or reduction.

Adsorption from solution—Adsorption is defined as the selective removal of metal salts or metal ion species from their solution by a process of either physisorption or chemical bonding with active sites on the support. Depending upon the strength of adsorption of the adsorbing species, the concentration of the active material through the support may be varied and controlled.

5 Pillared Clays

Many heterogeneous catalytic reactions are carried out at high temperatures of over 200 °C. As mentioned earlier natural clay minerals suffer from the disadvantage of interlayer collapse at temperatures greater than approximately 200 °C with consequent loss of catalytic activity.[11] Clays pillared with robust and large inorganic hydroxyl cations overcame the problem of thermal stability and attracted attention as potential catalysts.[12] Figure 3.2 is a schematic representation of pillared clay. Oxide pillars are inserted between the silicate layers, producing a heat-stable porous material.

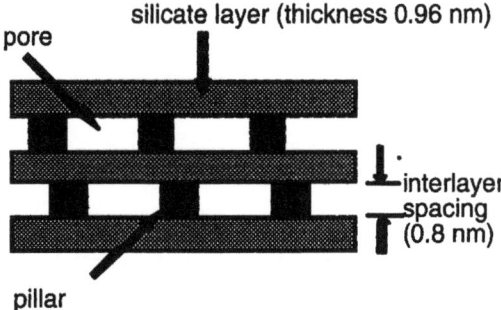

Figure 3.2 *A pillared clay*

The method for preparing an alumina-pillared clay is shown in Figure 3.3. Interlayer Na^+ ions in the original clay are exchanged with aluminium hydroxy cluster cations. When this material is calcined at between 300 and 500 °C, the cations become oxide pillars and release protons into the structure. An acidic and porous clay with a surface area of 200–500 $m^2 g^{-1}$ is thus formed.

The pillaring procedure makes good use of the intercalation ability of smectite clays and the resulting microporous solid retains a large exposed surface area even up to 600 °C.[13] Unfortunately, pillaring followed by calcination drastically reduces the CEC of the final material. This is a serious disadvantage of pillared clays compared with zeolites which have higher thermal stability and large CEC.

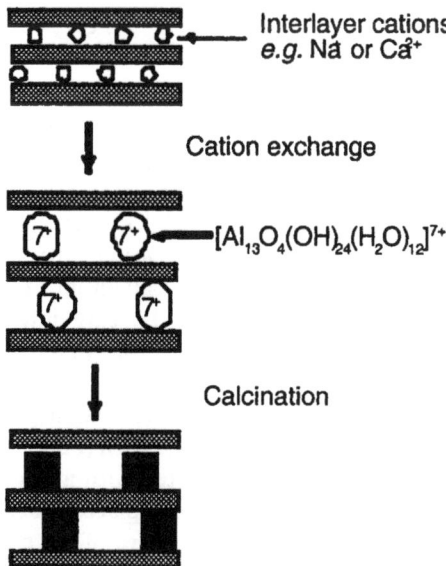

Figure 3.3 *Preparation of an alumina-pillared clay*

6   Clay Catalysed Reactions

Clays have been used as catalysts in a wide range of organic reactions. The most important and widely studied of these are described here.

Hydrogenation

Platinum particles have been incorporated into montmorillonite by aqueous exchange with $Pt(NH_3)_4^{2+}$ and subsequent reduction with hydrogen at 140 °C.[14] At loadings between 0.6 and 1.0% by weight of platinum a large proportion of metal particles are 3 nm in size. At higher loadings there is a tendency towards metal agglomeration to give metal films. It is interesting to note that the clay catalysts retain some swelling capacity, and are useful catalysts for condensed phase hydrogenations.

Palladium(II) ions have been shown to be effective hydrogenation catalysts when exchanged into a clay matrix. The most successful hydrogenation catalysts contain palladium complexes with large organic ligands. These ligands help keep the layers of the clay apart, conferring both size and shape selectivity. The rates of hydrogenation over a montmorillonite exchanged with diphenylphosphine palladium chloride decrease in the order cyclohexene > cyclopentene ≫ cycloheptene. The same catalyst also shows a high selectivity towards the hydrogenation of non-hindered double bonds (Reaction 2).[15]

$$(2)$$

Palladium can be introduced into montmorillonite hosts by a similar ionic exchange route to platinum using $Pd(NH_3)_4^{2+}$.[16] However, a higher level of dispersion can be achieved if the metal is complexed by more electrophilic ligands such as MeCN which enhance the reduction step. Crocker *et al.* have used the palladium complex $[Pd(PPh_3)(MeCN)_3]^{2+}$.[17] The palladium particles were mainly in the size range 2–4 nm. Despite the small metal particle size the measured surface area of supported palladium catalysts is rather low (*ca.* 30 m^2 g^{-1}) suggesting that access to the clay interlayers is blocked, presumably by large metal crystallites on the outer edges of the clay particles.

In an effort to reduce pore blocking, the same researchers added another step to the synthesis procedure. After ion exchange with $[Pd(PPh_3)(MeCN)_3]^{3+}$ the resultant palladium exchanged clay was treated with a sterically hindered and less electrophilic ligand $P(o\text{-}CH_3C_6H_4)_3$. This replaces MeCN ligands but only on accessible palladium ions on the outside edges of the clay platelets. Under mild reducing conditions the palladium(II) near the edges of the platelets remains largely unreduced. This results in no metal particles at the edges and access to the interlayer space is improved. The material shows a good degree of size selectivity with hydrogenation rates for cyclic alkenes decreasing in the order cyclopentene > cyclohexene $\gg$ cycloheptene.

Supported nickel catalysts are widely used for hydrogenation reactions. The most common supports are silica and activated carbon, however several studies have shown that clays are rather more effective in maintaining high metal surface area and can impart useful selectivity to the catalyst.[18] Deposition of nickel(II) by the incipient wetness technique [in which the volume of nickel(II) solution used is equal to the pore volume of the clay and then dried] followed by H_2 reduction was found to be the most effective method of generating a high dispersion of metal particles (< 10 nm in size) within the clay.

Fischer–Tropsch Synthesis

Ruthenium supported on pillared clay has been used for Fischer–Tropsch synthesis.[19] The catalyst is prepared by exchanging $HRu_3(CO)_{12}^+$ into alumina-pillared montmorillonite. Reduction with hydrogen at 400 °C yields ruthenium particles of less than 5 nm in diameter. The alumina pillars in this system are thought to be important because they provide hydroxyl groups to immobilise the metal centres, ensuring a high dispersion, with resultant high catalytic activity. The controlled interlayer space in the pillared material also provides catalyst selectivity. The catalyst shows a high selectivity towards branched chain alkenes, resulting from protonation of n-alkenes formed by Fischer–Tropsch chain propagation and subsequent rearrangement of the carbocations. Yields of branched chain hydrocarbons are comparable with the yields obtained using zeolites.

Pinnavaia *et al.* have shown that iron oxide pillared montmorillonite is an active catalyst for Fischer–Tropsch synthesis.[20] The interlayer spacing of the pillared product depended largely on the hydrolysis conditions used to generate the pillaring solutions. Under optimum hydrolysis conditions the pillared products contained 6.8–9.8 Fe^{3+} ions per $O_{20}(OH)_4$ unit cell and had an extremely large interlayer spacing of 1.6–1.9 nm. Reduction in hydrogen at 400 °C gave an active catalyst for CO hydrogenation at 275 °C and 120 psi. However, the iron oxide-pillared catalysts were found to age rapidly and this was manifested by the redistribution of iron to the edges of the clay particles resulting in a concomitant decrease in interlayer spacing.

Brönsted Acid Catalysed Reactions

Perhaps the best known catalytic property of clays is their Brönsted acidity, which as described in Section 3 can be controlled by ion exchange. The most widely studied organic reactions using clays as solid Brönsted acid catalysts are described in this section.

Addition, Elimination and Isomerisation

Many papers have been published on the catalytic activity of ion-exchanged clays. Ballantine *et al.* refluxed hex-1-ene with a Cu^{2+}-exchanged montmorillonite and produced a branched chain symmetrical ether (Reaction 3).[21] The clay acted both as a Brönsted acid catalyst as well as providing water molecules for the reaction.

$$R-CH=CH_2 \xrightarrow[+ H_2O]{H^+} R-\overset{+}{C}H-CH_3$$

(3)

The effect of varying the interlayer water content has been studied by the same authors using 2-methylpropene as a substrate. Two distinct pathways exist. With normal interlayer water content of approximately 12% the alkene adds to water. The same clay when thoroughly dehydrated (*i.e.* above 150 °C) catalyses the oligmerisation of 2-methylpropene.[22]

The Cu^{2+} montmorillonite was also found to be an effective catalyst for the addition of carboxylic acids to alkenes to give esters. Al^{3+}-exchanged montmorillonite was shown to catalyse the reaction between ethanol and hex-1-ene producing 2- and 3-ethoxyhexanes.

The montmorillonite catalysed hydration of ethene has been studied by Atkins *et al.*[23] The clay catalysts were prepared by ion exchange with various metal ions and by acid treatment with cold 5 mol dm^{-3} sulfuric acid. The ethene was mixed with water vapour before passing though a fixed catalyst bed held at a temperature between 200 and 300 °C. The order of reactivity was found to be $Al^{3+} \gg Fe^{3+} > Cr^{3+} > H^+$. The surprisingly low activity of the H^+ montmorillonite was explained by X-ray diffraction measurements which showed that the H^+ clay had collapsed under the reaction conditions with a decrease of the interlayer spacing from 1.5 nm to 0.96 nm.

Hydrocarbon Cracking

Gas oil cracking has been catalysed by pillared clays which gave high activity and selective yields of gasoline under moderate conditions. The activity and selectivity are affected by the type of host clay; for example, hectorite exhibits a high gasoline selectivity and minimises light gas production, although it is not as catalytically active as montmorillonite. Recently, pillared acid activated clays (PAAC), in which the clay is acid treated before pillaring, have been shown to be more active than conventional pillared clays for catalytic cracking.[24] The PAAC catalysts have interlayer spacings, surface areas and thermal stability comparable to conventional pillared clays but possess significantly higher pore volume and Brönsted acidity. Additionally the pore size distribution of the PAAC catalyst is broader and shifted toward larger pores, with the extent of the shift dependent on the severity of acid treatment of the original clay.

The largest barrier to using pillared clays as industrial cracking catalysts is deactivation due to coke formation, which is an inevitable by-product of shape-selective acid-catalysed organic conversions. The formation of coke can be reduced to an extent by adding a small amount of Pd to the pillared clay.

Friedel–Crafts Alkylation

Laszlo and co-workers have reported that K10 montmorillonites ion-exchanged with various transition metal cations are efficient catalysts for Friedel–Crafts alkylation of aromatics with a wide range of alkylating agents such as alcohols, alkenes and alkyl halides.[25]

Following on from the seminal work of Laszlo and co-workers, Mayoral *et al.* made catalytic activity measurements on transition and main group metal exchanged K10 in the alkylation reaction between anisole and dienes.[26] The samples exchanged with H^+, Al^{3+}, La^{3+} and Fe^{3+} gave a large amount of monoalkylated product initially but the yield decreased with time because of retro-alkylation followed by diene polymerisation. The use of clays which had been thoroughly dehydrated at 550 °C before use strongly reduced these side reactions. Calcination under these conditions eliminates most of the Brönsted acid sites which are thought to be responsible for diene polymerisation and retro-alkylation.

The enhancement in catalytic activity of cations such as Zn(II) which has been achieved both through ion exchange as well as deposition of Zn(II) salts onto clay surfaces led to studies of the acidity and catalytic activity of such ions when incorporated directly into the lattice sites of synthetic clay minerals. Luca *et al.* showed that Lewis acid sites are generated on Zn^{2+}-substituted fluorohectorite.[27] The Zn^{2+}-substituted fluorohectorite was synthesised by a sol-gel route. The sol was allowed to crystallise in a Parr autoclave at 250 °C for 24 hours. The Lewis acid sites were identified as Zn^{2+} at the edges of the fluorohectorite crystallites and were active towards the Friedel–Crafts alkylation of benzene with benzyl chloride.

One of the important catalytic processes based on shape selectivity is the alkylation of biphenyl with propene (Reaction 4). Pinnavaia *et al.* have shown that mesoporous clays such as K10 and alumina-pillared montmorillonite are more selective than homogeneous acid catalysts, although not as catalytically active.[28] The product distribution obtained with sulfuric acid was 44% *ortho*, 24% *meta* and 32% *para*.

(4)

The heterogeneous catalysts tested produced smaller amounts of the *ortho* isomer, the pillared montmorillonite being the most selective in this respect, giving only 15% of *ortho* product. The catalysts with a lower proportion of micropores such as delaminated synthetic hectorite and K10 gave lower yields of *meta* and *para* isomers than the microporous pillared clays. The explanation for these lower yields is that the *ortho* isomer diffusion, being slower than that of the *meta* and *para* isomers, isomerises or undergoes a second alkylation more readily than the other isomers. With catalysts containing a large proportion of micropores, diffusion control becomes important and the yield of the *ortho* isomer depends on the surface area contained in the micropores. This was confirmed by the fact that the yield of *ortho* isomer decreased as the particle size was increased.

There are a number of commercially acid-treated clays and although K10 (available from Fluka) has been very popular for alkylation reactions, other acid-treated clays have been found to be equally effective in certain cases. For example, the acid-treated clay Engelhard F-24 was found to be a very effective catalyst for the alkylation of diphenylamine with α-methylstyrene (Reaction 5).[29] The dialkylated diphenylamines produced in this reaction are industrially important as antioxidants and heat stabilisers in polypropylene and polyethers.

(5)

Natural clays have also been used as catalysts for Friedel–Crafts alkylation reactions.

Okado *et al.* reported on the use of natural vermiculite as a highly active and selective catalyst for the Friedel–Crafts alkylation reaction between benzene and 4-methylbenzyl chloride.[30] The important feature of vermiculite for this reaction appears to be its unusually high structural iron content. Indeed, a vermiculite containing a much smaller amount of iron exhibited a lower activity than H^+ montmorillonite.

Results from powder X-ray diffraction measurements indicated that the interlayer spacing of vermiculite did not change during the reaction. This implies that the interlayer region is too narrow to accommodate the reactants and only Lewis acid sites on the external surfaces, probably Fe^{3+} cations, are responsible for catalysing the reaction.

Salmon *et al.* used a natural montmorillonite to catalyse the intermolecular condensation of toluene in the presence of bromine to produce *ortho-* and *para-*phenyltolylmethanes.[31] The reaction was carried out at reflux using carbon disulfide as a solvent. The first stage of the reaction is believed to be the generation of benzyl bromide. A bromide ion is abstracted from benzyl bromide by a Lewis acid site on the clay surface. The resulting electrophilic benzyl cation then attacks toluene.

It has been found that Al^{3+}-exchanged natural saponite prepared by cation exchange with $Al(NO_3)_3$ exhibits a high catalytic activity towards the alkylation of toluene by methanol.[32] The Al^{3+} saponite retained a high surface area even after calcination at 400 °C and this is due to its unique three-dimensional structure. Saponites tend to form the so-called 'house-of-cards' structure shown in Figure 3.4, due to the electrostatic interaction between negatively charged faces and positively charged edges of the clay platelets.[33]

The X-ray diffraction pattern of saponite exhibits a relatively weak and broad (001) reflection compared with montmorillonite, indicating a lack of long-range layer ordering. This characteristic of saponite can be ascribed to the three-dimensional voluminous house-of-cards structure. The house-of-cards structure also contributes to the high catalytic activity in comparison to montmorillonites, which do not form the house-of-cards structure. Alkylation activity

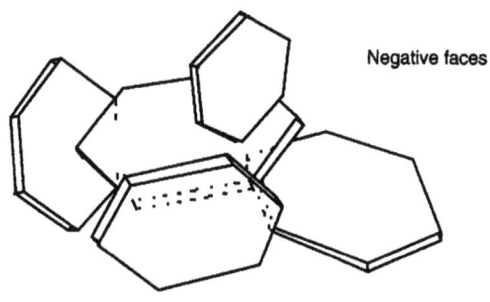

Negative faces

Positive edges

Figure 3.4 *House-of-cards structure of saponite*

depends directly on the type of exchanged cation. The saponites exchanged with Zr, Al or Ti cations, which have large hydration energies, exhibited very high activities comparable to that of alumina-pillared montmorillonite.

The use of clay-based supported reagent catalysts in Friedel–Crafts reactions adds a new dimension to this area and has resulted in successful application on an industrial scale. This is discussed in Chapter 4.

Aldol Condensation

There has been a quite limited number of reports of clay-catalysed aldol condensations. One of the more interesting of these is the aluminium-exchanged montmorillonite (Al^{3+}-mont) catalysed cross-aldol reaction of silyl enol ethers with aldehydes (Reaction 6).[34] The reaction proceeded smoothly under mild conditions to give the corresponding aldol adduct in good yield.

$$
\underset{R_1}{\overset{OSiMe_3}{\diagup}}\!\!\!\diagdown\!\!R_2 \;+\; R_3CHO \;\xrightarrow{\text{clay catalyst}}\; \underset{R_1}{\overset{O}{\diagup}}\!\!\!\underset{R_2}{\diagdown}\!\!\!\underset{R_3}{\overset{OSiMe_3}{\diagup}} \tag{6}
$$

The diastereoselectivity of the reaction was independent of the catalyst but was affected by the nature of the solvent. The *threo* isomer was preferentially formed in toluene, while the *erythro* isomer was formed in 1,2-dimethoxy-ethane. The proton-exchanged montmorillonite (H^+-mont) showed similar activity and diasteroselectivity to Al^{3+}-mont. This fact suggests that the exchangeable Al^{3+} cations in the montmorillonite do not function as Lewis acid sites and it is the Brönsted acid sites that are essential for catalysis of the aldol reaction.

Oxidation

Mn^{2+}-exchanged montmorillonite has been found to be an effective catalyst for the oxidation of alkanes such as cyclohexane, cyclooctane and adamantane to give the corresponding ketones (Reaction 7). Tateiwa *et al.* prepared a Mn^{2+}-exchanged clay catalyst by treatment of a commercially available Na^+-montmorillonite with $Mn(NO_3)_2 \cdot 6H_2O$ in aqueous solution at 50 °C for 24 hours.[35] The resulting solid was collected by filtration, washed with water and dried in air at 120 °C. The optimum conditions for the oxidation of cyclohexane were found to be 70% *tert*-butyl hydroperoxide at 60 °C in the presence of Mn^{2+}-exchanged clay suspended in benzene. It is interesting to note that the addition of 4A molecular sieves to the reaction increased the product yields and it was thought that this may act as a dehydrating agent keeping the system dry and the catalyst active. Other ion-exchanged montmorillonites such as Al^{3+}-,

Fe^{3+}-, H$^+$- and Na$^+$-exchanged montmorillonite showed only very low catalytic activity towards the oxidation.

(7)

7   New Developments in the Context of Clean Synthesis

Clays and clay-based supported reagents (see Chapter 4) have become established catalysts in organic synthesis and it is not suprising that they have also become commonly associated with the clean synthesis of organic compounds. Some interesting recent examples are described below.

The polymer industry requires large quantities of 4,4'-diaminodiphenyl-methanes. These compounds are used in various applications as polymer additives. Currently they are manufactured by the acid-catalysed condensation of anilines with formaldehyde but like so many reactions of this type, the processes require the neutralisation of waste acid leading to large amounts of salt waste. Cleaner processes are required and solid acids offer a way forward. One potentially useful example of this is the aqueous reaction of aromatic anilines adsorbed on kaolinite with formaldehyde which cleanly give the desired products (Figure 3.5).[36] In a typical experiment, the clay is stirred with water before aniline is added to the stirred solution. To this is slowly added formaldehyde solution. The precipitated 4,4'-diaminodiphenylmethane can be extracted in hot ethanol in a yield of 96%. Similarly, several substituted analogues have been prepared in yields of 70–99%.

A clean method for the rapid and solventless synthesis of herbicides based on 2,4-dichlorophenoxyacetic acid (2,4-D esters) has been described in which

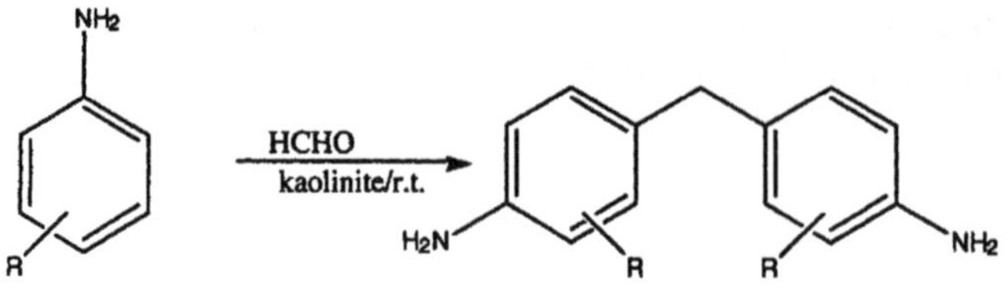

R = H, Me, Cl, CO$_2$H   Yields = 68 (2,3-dimethylaniline) to 99%

Figure 3.5   *Condensation of anilines with formaldehyde using aqueous clay*

micro-particulate inorganic materials such as clays act as catalysts (Figure 3.6).[37]

The reaction is strongly accelerated when subject to microwave radiation, a popular technique for clay-catalysed reactions.[38] These esters are extensively used around the globe as effective hormonal herbicides with high selectivity in applications to crops including cereals, grazing land and sugar cane plantations. They are also used domestically in gardens in the USA and Western Europe. Current manufacturing methods are environmentally hazardous.[37] The clay-based methodology offers another interesting advantage. Large amounts of clay minerals are used as pesticide carriers (around 200 000 tonnes per annum both in the USA and in Western Europe). The 2,4-D prepared on clay could be applied as is, *i.e.* on the mineral used to catalyse its formation. To further support this concept, it has been shown that agricultural soils from different parts of the world can be used to catalyse the esterification reaction after

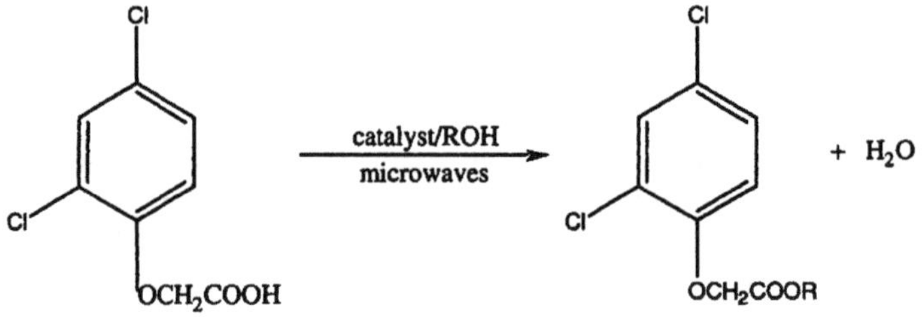

2,4-D acid   2,4-D esters

Catalyst	Ester yield for R = i-Pr (%)
Kaolinite	47
Saponite	100
Sepiolite	97

Figure 3.6   *Formation of 2,4-D esters using clay catalysis*

thermal activation. This idea could be applicable to other agrochemical products amenable to clean synthesis methods based on clay minerals.

Two important processes for the construction of molecules, conjugate addition and alkoxyalkylation, can be carried out using commercially available and reusable clays.[39] The products of such reactions are industrially important. For example, alkoxyalkylation leads to intermediates in the production of quinolines, which have value as bioactive compounds. Addition reactions have widespread applications in commercial processes. They are likely to be increasingly favoured over alternatives as they can be 100% atom efficient. The reaction conditions using clay catalysts required are mild, and relatively benign hydrocarbon solvents are employed. Reaction selectivities are also good. The use of clays in Michael reactions may be considered as examples of reactions employing the dual catalytic nature of the materials. Montmorillonite clays are crystalline aluminosilicates, whose multilayered structure is characterised by the presence of Lewis acidic sites which are located mostly on the edges of the layers, capable of activating Michael reaction donors, and Brönsted acid sites which are mainly in the interlamellar region and which can activate Michael reaction acceptors (Figure 3.7). Thus ethyl benzoyl acetate reacts readily with methyl vinyl ketone in the presence of acid-treated clays at room temperature in a solvent such as hexane. The catalyst can be used several times with little loss in activity.

The alkoxyalkylation reaction of carbonyl compounds can be considered as an orientated cross-aldol condensation between two masked carbonyl compounds, an acetal and a silyl enol ether. The key step involves the formation of an electrophilic species by reaction of the acetal with catalytic amounts of a Lewis acid and the right catalyst can lead to excellent diastereo- and enantio-selectivities. Clays are satisfactory catalysts in these reactions, with the acid-treated clay K10 performing better than the more powerfully acidic clay KSF,

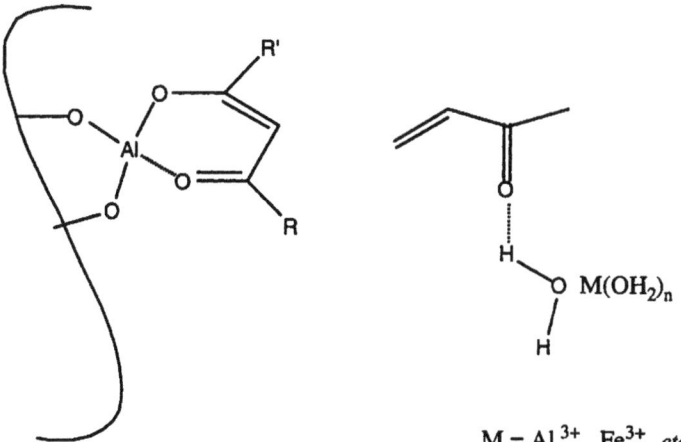

$$M = Al^{3+}, Fe^{3+}, etc.$$

Figure 3.7 *Simultaneous activation of the Michael donor and acceptor by a clay catalyst*

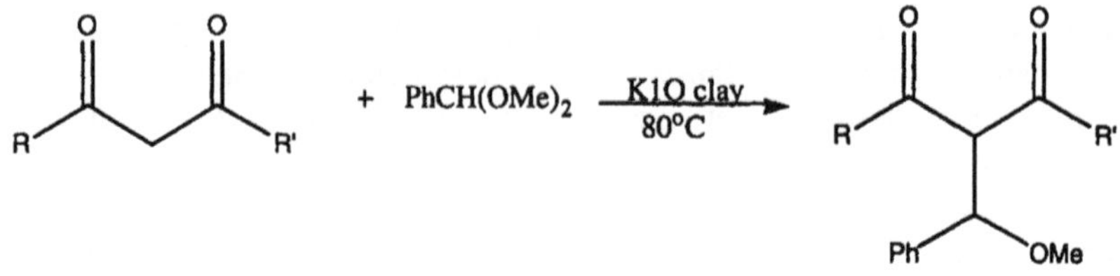

Yield (%)

R = Ph	R' = OEt	86
R = Ph	R' = Me	42
R = i-Pr	R' = OEt	43% E- and 40% Z-isomers

Figure 3.8 *Clay-catalysed alkoxyalkylations*

presumably because the latter causes decomposition of the acetal (Figure 3.8). In reactions employing new stereogenic centres, however, a poor degree of stereoselectivity was observed and the composition of the diastereoisomeric mixtures varied considerably after the purification procedure presumably because of the ready enolisation of the 1,3-dicarbonyl product.[39]

Triphasic systems involving clays pillared with surfactants as catalysts offer versatile routes to a range of useful benzylic compounds.[40] By using the organo-clay assemblies with sodium cyanide, thiocyanate and hydroxide, it is possible to prepare benzonitriles, benzyl thiocyanates and benzyl alcohols from the corresponding benzyl chlorides. Most of these reactions occur in yields of *ca.* 80%.

While clays are commonly associated with acid-catalysed reactions, their environmental acceptability, good availability and ease of use makes them attractive materials for other applications. Some clay materials can be used as solid bases. These include magnesium hydrotalcites, which can be used, for example, in the synthesis of coumarins, which are important biologically active materials for which cleaner synthetic methods are required.[41] The coumarins are synthesised *via* Knoevenagel condensation of various phenols with 2-substituted ethyl acetates in hot toluene. Product yields are generally good when using Mg–Al hydrotalcite although other solid bases including alumina and CsX-zeolite give no reaction. The Henry reaction is an important class of C–C bond forming reaction catalysed by bases. It gives nitroalkanols, which are important intermediates for various useful compounds such as amino alcohols. Difficulties that give rise to waste include dehydration and Cannizzaro reaction of the aldehyde component. Mg–Al hydrotalcites seem to overcome these problems giving selective and clean reactions under mild conditions.

The mechanism of the reaction can be assumed to involve the abstraction of a proton from the active methylene group of the nitro compound which gives a carbanion that can be stabilised by the Al sites in the lattice of the hydrotalcite.

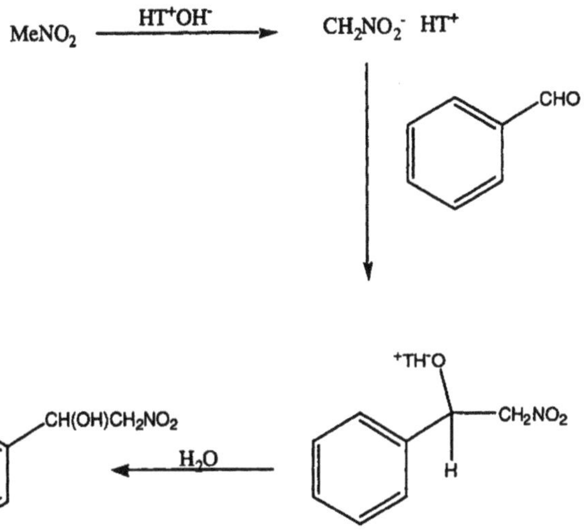

Figure 3.9 *Henry reaction catalysed by Mg–Al hydrotalcite*

This carbanion further adds to the carbonyl compound to form an intermediate, which removes a proton from water to give the final nitro aldol product (Figure 3.9). Solid bases without Brönsted hydroxy groups such as decarbonated hydrotalcites are not active in this reaction. The Mg–Al hydrotalcite catalyst is non-toxic and can be recycled with little loss in activity.[42]

References

1 J.H. Purnell, J.M. Thomas, P. Diddams, J.A. Ballantine and W. Jones, *Catal. Lett.*, 1989, **2**, 125.
2 R. Gregory and D.J. Westlake, *EP Appl.* 0083970/1983.
3 'Preparative Chemistry Using Supported Reagents', ed. P. Laszlo, Academic Press, San Diego, 1987.
4 *International Symposium; Supported Reagent Chemistry*, Royal Society of Chemistry, University of York, 2–5 July, 1991.
5 L. Pauling, *Proc. Nat. Acad. Sci. USA*, 1930, **16**, 123.
6 R.E. Grim, 'Clay Mineralogy', McGraw-Hill, New York, 1953.
7 J.M. Thomas and J. Klinowski, *Adv. Catal.*, 1985, **33**, 335.
8 M.M. Mortland and K.V. Raman, *Clays and Clay Minerals*, 1968, **16**, 393.
9 A.G. Pickett and M.M. Lemcoe, *J. Geophys. Res.*, 1959, **64**, 1579.
10 E.J. Houdry, W.F. Burt, E.A. Pew and W.A. Peters, *Natl. Petroleum News*, 1938, **30**, 570.
11 E.B. Cornelius, J. Holmes and G.A. Mills, *J. Phys. Colloid Chem.*, 1950, **54**, 1170.
12 H. Sakurai, K. Urabe and Y. Izumi, *Catalyst*, 1988, **28**, 397.
13 T.J. Pinnavaia, *Science*, 1983, **220**, 365.
14 J.B. Harrison, V.E. Berkheiser and G.W. Erdos, *J. Catal.*, 1988, **112**, 126
15 B.M. Choudary and K.K. Rao, *Tetrahedron Lett.*, 1992, **33**, 121.
16 T. Matsuda, T. Fuse and E. Kikuchi, *J. Catal.*, 1987, **106**, 38.
17 M. Crocker, R.H.M. Herold, J.G. Buglass and P. Companje, *J. Catal.*, 1993, **141**, 700.

18 M. Patel, *Clays and Clay Minerals*, 1982, **30**, 397.
19 E.P. Giannelis, E.G. Rightor and T.J. Pinnavaia, *J. Am. Chem. Soc.*, 1988, **110**, 3880.
20 T.J. Pinnavaia, E.G. Rightor and M. Tzou, *J. Catal.*, 1991, **130**, 29.
21 J.A. Ballantine, J.H. Purnell and J.M Thomas, *Clay Miner.*, 1983, **18**, 347.
22 J.A. Ballantine, W. Jones, J.H. Purnell, D.T.B. Tennakoon and J.M. Thomas, *Chem. Lett.*, 1985, 763.
23 M.P. Atkins, D.J.H. Smith and D.J. Westlake, *Clay Miner.*, 1983, **18**, 423.
24 R. Mokaya, W. Jones, Z. Luan, M.D. Alba and J. Klinowski, *Catal. Lett.*, 1996, **37**, 113.
25 P. Laszlo and A. Mathy, *Helv. Chim. Acta*, 1987, **70**, 577.
26 J.A. Mayoral, C. Cativiela, J.I. Garcia, M. Garcia-Matres, F. Figueras, J.M. Fraile, T. Cseri and B. Chiche, *Appl. Catal.*, 1995, **123**, 273.
27 V. Luca, L. Kevan, C.N. Rhodes and D.R. Brown, *Clay Miner.*, 1992, **27**, 515.
28 T.J. Pinnavaia, J. Butruille, L.J. Michot and O. Barres, *J. Catal.*, 1993, **139**, 664.
29 S. Chitnis and M. Sharma, *J. Catal.*, 1996, **160**, 84.
30 S. Okado, K. Tanaka, Y. Nakadaira and N. Nakagawa, *Bull. Chem. Soc. Jpn.*, 1992, **65**, 2833.
31 M. Salmon, E. Angeles and R. Miranda, *J. Chem. Soc., Chem. Commun.*, 1990, 1188.
32 H. Sakuria, K. Urabe and Y. Izumi, *Shokubai, (Catalyst)*, 1986, **28**, 397.
33 H. van Olphen, 'Introduction to Clay Colloid Chemistry', Wiley, New York, 1977.
34 M. Arai, S.L. Guo, M. Shirai, Y. Nishiyama and K. Torii, *J. Catal.*, 1996, **161**, 704.
35 J. Tateiwa, H. Horiuchi and S. Uemura, *J. Chem. Soc., Chem. Commun.*, 1994, 2567.
36 D. Bahulayan, R. Sukumar, K. Raghavanpillai and M. Lalithambika, *Green Chemistry*, 1999, 191.
37 L. Lami, B. Casal, L. Cuadra, J. Merino, A. Alvaraz and E. Ruiz-Hitzky, *Green Chemistry*, 1999, 199.
38 'Microwave Enhanced Chemistry. Fundamentals, Sample Preparation and Applications', eds. H.M. Kingston and S.J. Haswell, American Chemical Society, Washington, 1997.
39 A. Soriente, R. Arienzo, M. De Rosa, L. Palombi, A. Spinella and A. Scettri, *Green Chemistry*, 1999, 157.
40 R.S. Varma, K. Pitchumani and K.P. Naicker, *Green Chemistry*, 1999, 95.
41 A. Ramani, B.M. Chanda, S. Velu and S. Sivasanker, *Green Chemistry*, 1999, 163.
42 B.M. Choudary, M. Lakshmi Kantam, Ch. Venkat Reddy, K. Koteswara Rao and F. Figueras, *Green Chemistry*, 1999, 187.

CHAPTER 4

Supported Reagents

1 Introduction to Supported Reagent Chemistry

The term 'supported reagent' has been used to describe a wide range of materials involving an inorganic or organic support onto which a 'reagent' species has been chemically or physically adsorbed. The concept can be traced back to a 1924 report[1] on the use of chemical reagents on porous carriers to achieve some 'highly selective and mild chemical reactions'. Many regard the 1968 paper by Fetizon as a landmark article.[2] In this, a method was described whereby the very inactive solid oxidant silver carbonate was rendered active in some organic oxidations by supporting it on celite. Further landmarks are the first reviews on the use of inorganic-supported reagents in organic chemistry in 1978[3] and 1979,[4] the major text on polymer-supported reagents published in 1980,[5] the first book on supported reagents published in 1987,[6] the first international symposium on supported reagents in 1991 and the first reported industrial applications of inorganic support-based materials in the 1990s.[7]

Some of the key advantages of supported reagents compared to the unsupported reagent are:

- good dispersion of active sites and the concentration of sites within small pores can lead to significant improvements in activity
- the presence of molecular dimension pores and the adsorption of reactant molecules on the material surface can lead to improvements in reaction selectivity
- easier and safer handling

These, coupled with the normal advantages of solid catalysts of easier recovery and reuse, make supported reagent type catalysts attractive materials to organic and industrial chemists. It should be borne in mind, however, that there are not insignificant drawbacks with their use, notably the added cost and weight of the support and the possibility of trapping reactant and product molecules on the material (including side-products which can accumulate leading to loss of activity).

2 Types of Supported Reagents

Whereas the motivation behind Fetizon's original work and much of the early research on supported reagents was directed towards stoichiometric reagents such as silver carbonate (so that at best only the support could be recovered unchanged), much of the more recent published work has dealt with genuinely catalytic materials whereby the entire supported reagent can, at least in principle, be recovered unchanged from the reaction. The main emphasis here will be on catalytic materials.

One sub-classification of supported reagents is that between organic and inorganic supports. The former are polymeric, such as cross-linked polystyrenes, and are numerous, covering ion-exchange resins through to those based on quite complex surface architectures. The polymer may be a linear soluble species of various molecular weights, a cross-linked network, a membrane or even a macroscopic polymeric object.[8] Soluble polymers are of limited value because they lack one of the key advantages of solid catalysts, that of easy separation from the reaction medium, although micro- and ultrafiltration techniques may alleviate the problem. Macroscopic object supports are very important in medical diagnostic kits but are not really viable for study with conventional laboratory reactors. This may change as novel process engineering brings in more complex reactors such as those with integrated catalytic plates. The majority of studies on polymer-supported reactions have been performed using cross-linked polymers, typically in the form of suspension resin beads of 10 µm to 1 mm in diameter, which are effectively insoluble in most solvents. The emphasis of this chapter will be on inorganic supported reagents.

There are numerous inorganic supports that are potentially suitable for conversion into supported reagents. These include silicas, aluminas, carbons (notably charcoals), montmorillonites, zeolites, and other aluminosilicates, as well as more complex materials such as partially substituted aluminsilicates (*e.g.* aluminophosphates or ALPOs) and more complex materials such as heteropolyacids. All of these materials can exist in high surface area forms (100–1000 m^2 g^{-1}) and are normally porous, with average pore diameters ranging from the microporous zeolites (from 0.3 nm) to some macroporous silicas (up to 100 nm at least). The particle sizes of such materials can range from coarse (in the case of formed materials) to very fine (< 1 µm). The materials can be sub-classified in terms of their being: crystalline with regular pore structure and a very narrow pore size distribution (*e.g.* zeolites); amorphous with a possible wide range of pore structures and a broad pore size distribution that can range over several tens of nanometres (*e.g.* conventional silica gels); and flexible layered structures (*e.g.* montmorillonite clays). The range of suitable materials available has been greatly enhanced through the discovery of the hexagonal mesoporous silicas (HMSs),[9] notably the MCM materials developed by Mobil.[10] These offer the intriguing possibility of being able to grow silicas, aluminosilicates and potentially many other materials to almost any pore size. Such materials commonly have very high surface areas (> 1000 m^2 g^{-1} is not uncommon). The list can be extended still further if lower surface area materials

Table 4.1 *Inorganic support materials*

Support	Comments
Silica gels (commercial)	Widely available and inexpensive; high surface area, mesoporous and normally broad pore size distribution; surfaces are heavily hydroxylated and easily functionalised (halogenation, silylation)
Structured silicas (synthetic)	MCMs and HMS materials are mesoporous; prepared using sol-gel methods using onium or amine templates, very high surface areas ($> 1000 \ m^2 \ g^{-1}$) and narrow pore size distribution but little long-range order; often less hydrophilic than normal silicas
Montmorillonites (natural and synthetic)	Natural clays can have swelling structure giving microporosity; pillared clays have larger pores; acid-treated clays are partially mesoporous (and may change with ageing)
Aluminas	Moderate surface areas; available in acidic, basic and neutral forms
Zeolites	Microporous and with tight pore size distribution; high surface areas; highly structured; small pores can give high degree of shape selectivity but may cause diffusional problems, especially in liquid-phase systems
Other materials	Charcoals can have very high surface areas ($> 1000 \ m^2 \ g^{-1}$) but complex surface functionalities; minerals such as calcium fluoride have low surface areas but can be useful as highly inert surfaces

are included, the only sensible limitation being that the support is insoluble in typical reaction media and even a highly inert and apparently surface inactive solid such as calcium fluoride has found some use as a support.[11,12]

The choice of support material for the preparation of a supported reagent can be the most difficult step in the entire operation. Several important factors are described below and are summarised in Table 4.1.

Porosity

Microporous supports, notably the zeolites, offer the valuable possibility of shape selectivity brought about by the molecular scale pores restricting the size and shape of substrate and product molecules as well as transition states. Unfortunately, the slow diffusion of molecules in the liquid phase through molecular scale channels can result in very slow reaction rates and their use as catalysts is more commonly associated with gas-phase reactions. The effect can be exaggerated by surface modification or especially by the inclusion of ionic catalytic centres which makes the internal structure extremely polar. Since organic reaction media are often quite non-polar, the partitioning of a low-polarity substrate between the bulk liquid and the pore structure of the catalyst is very unfavourable for reaction. This has to be countered by the fact that such a highly polar region within the catalyst can give extremely favourable activation effects through the high local reaction fields. There are some successful liquid-phase applications for microporous solids, typically involving smaller molecules, including various condensation reactions which produce small molecules for which zeolitic materials have a high affinity (especially water). Other promising new application areas include the use of a zeolite-bound catalysts as enzyme mimics for synthesising chiral drugs and agrochemicals.[13] Rather larger pore materials include the aluminophosphates and pillared clays (layered montmorillonite clays with chemical props fixing the structure with interlayer gaps of about 1–2 nm). More commonly, catalysts designed for liquid-phase applications are based on mesoporous materials such as silica gels, acid-treated clays and HMS-type materials with average pore diameters ranging over 2–10 nm. These are mostly open enough to allow good molecular diffusion rates even for quite large molecules in the liquid phase. Generally any improvements in reaction selectivity when using these materials can be attributed to adsorption effects or to differences in molecular diffusion rates through the (normally polar) pores. However, the smaller controlled pore HMS materials have been shown to have small enough pores to give some apparent shape selectivity with larger substrate and product molecules.[14] The other advantage of the HMS materials is their narrow pore size distribution. Commercial support materials such as silica gels have very broad pore size distributions. While in most cases this may not be a matter of great importance with regard to reaction selectivity, the smaller pores may be largely wasted due to poor access and/or easy poisoning. It is also worth noting that on preparing a supported reagent from such materials, the larger pores will fill first and narrower channels and pore structures may present few active sites if the catalyst has not been carefully prepared.

Chemical Composition

The actual chemical nature of the support material may be and often is of direct importance to its usefulness as a support material.[15] Silicas can react with small nucleophiles such as F^-, OH^- and CN^-. Thus, silica-supported fluorides are inactive, both as nucleophilic fluorinating agents and as bases. Similarly, silicas are not effective support materials for cyanides due to the formation of strong Si–CN bonds. For different reasons, an acidic clay would not be a suitable support for cyanides, due to the possible formation of toxic HCN. Charcoal is the most effective support material for stabilising Cu(I), probably due to its aromatic character.[16] For many chemisorbed supported reagent catalysts, silicas are preferred since they give relatively strong surface bonds. However, Si–O–C bonds are hydrolytically vunerable and direct Si–C bonds are preferred.[17]

Surface

High surface areas are generally preferred since this should maximise the number of active sites and increase substrate adsorption, as well as increase the contribution from heterogeneous chemistry in cases where both homogeneous and heterogeneous reactions can occur.[15] The highest surface area supports are the HMS type materials; commercial silica gels and zeolites have high surface areas, with clays and aluminas generally having lower values. Approximately 80% of the surface area of most support materials is inside the pores. On filling the pores with a polar reagent, the experimental surface areas are generally reduced, since the non-polar adsorbate (typically N_2) is repelled from the more highly polar regions of the support. Removal of the reagent can fully restore the apparent surface. Overloaded supported reagents can have excess reagent physically blocking the entrance to pores.

Commonly used support materials used under normal conditions have more or less fully hydroxylated surfaces with a small number of Si–O–Si siloxane bridges which can be converted into SiOH groups by pretreatment with aqueous HCl. Materials prepared *via* sol-gel methods from aqueous alcohol are likely to contain surface alkoxy groups.[18] The hydroxylated surfaces of silicas and other support materials are highly polar. Values of polarities/polarisibilities (measured *via* the non-specific solvatochromic parameter, π^*) higher than any known liquids have been estimated[19] but a better guide might be the relative values of the general solvatochromic parameter E_T^N which show that the unfunctionalised hydroxylated surface of a typical silica is considerably more polar than organically modified surfaces, even those containing polar functional groups (Table 4.2).[20] The relative decrease in polarity per mmol of organic groups per gram of silica (ΔE_T^N) is broadly in agreement with the relative E_T^N values found for comparable liquid compounds (from TMS = 0 to propionitrile = 0.401).

Surface hydroxyls help to bind molecules (substrates and physisorbed reagents) onto surfaces and hence encourage reaction and inhibit leaching. Unfortunately, they can also bind product molecules creating difficulties in catalytic turnover and in site poisoning. Surface hydroxyls might be

Table 4.2 *Polarities of a range of organically modified silicas measured via the use of Reichardt's dye*

Material	Degree of functionalisation[a] (mmol g^{-1})	λ_{max} (nm)	E_T^N	ΔE_T^N (mmol g^{-1})
silica-OH	—	477	0.967	—
silica-OSi(CH$_3$)$_3$	0.53	560	0.630	−0.634
silica-Ph	1.00	610	0.500	−0.467
silica-(CH$_2$)$_3$Cl	0.64	554	0.645	−0.503
silica-(CH$_2$)$_3$NH$_2$	0.99	590	0.546	−0.425
silica-(CH$_2$)$_2$CN	0.99	557	0.636	−0.334
silica-(CH$_2$)$_3$N⟨⟩=N	0.97	611	0.497	−0.485
polystyrene	—	752	0.225	—

[a] Measured by elemental analysis.

disadvantageous to the reagent, *e.g.* through the formation of strong bonds such as the strong OHF$^-$ hydrogen bond with supported fluorides, which effectively kill the nucleophilicity of the F$^-$ (although the materials are useful general bases).[21] For supported reagents with chemically bonded active sites, the hydroxyl group is the anchor onto which the catalytic site is built, but it is worth noting that full functionalisation of surface hydroxyls is not realised in practice; indeed, most workers in the area would willingly settle for a 50% success rate. This means that surface hydroxyls are almost inevitably left on the surface, potentially causing problems in terms of competitive adsorption or reactions with the exposed surface. Here, surface silylation may help to reduce the excess hydroxyl population although it is an expensive and never fully efficient treatment.

3 Preparation of Supported Reagents

There are numerous methods for preparing supported reagents and they are summarised in Table 4.3. Impregnation is an extremely versatile technique (although it is not applicable to insoluble reagents) which can be controlled to give good dispersion and a known loading of reagent. It has been successfully applied to many of the catalysts described later in this chapter including supported zinc halides (used in reactions including Friedel–Crafts alkylations and bromination reactions) and supported fluorides (used as solid bases).[17,21] It is also used extensively in the industrial scale manufacture of solid catalysts.[7]

Table 4.3 *Methods of preparing supported reagents*

Method	Comments
Sol-gel synthesis	Direct method; can use to introduce functional group, *e.g.* *via* (MeO)$_3$Si–R; resulting materials can have high surface areas; may be difficult to control; becoming more popular
Post-modification of support material	Most commonly using silane, but may be less stable than analogous sol-gel materials due to only partial surface reaction; alternative methods include *via* chlorination (*e.g.* Si–Cl (Si–R)
Impregnation (pore filling followed by evaporation of solvent)	Requires a suitable solvent; enables good control over dispersion and loading; likely to give physisorbed reagent only
Ion exchange	Simple and effective where appropriate
Precipitation/ co-precipitation	Useful for poorly soluble reagents but may be difficult to control
Adsorption from solution	Easy to carry out but loading may be low
Mixing/grinding	Easy and avoids other chemicals, but unlikely to give good dispersion

Ion exchange is the most important technique for the preparation of clay and zeolite catalysts from the preformed supports. It is used to make several catalysts described here, including montmorillonite-Fe^{3+}, which is a useful Diels–Alder catalyst that can function in water and zeolite-Na^+,H^+ (*i.e.* partially proton-exchanged zeolite), which can catalyse some aromatic chlorinations with shape selectivity.[15,21,22]

The sol-gel technique is essential for the preparation of catalysts not involving preformed supports.[17,23–25] It gives more potential flexibility than any other since the bulk as well as surface composition and structure are adjustable. The method can be used to prepare the support or the entire catalyst, typically by incorporating an organic functionality into the gel. The resulting materials can have high surface areas with strongly bonded sites at remarkably high loadings (sometimes as high as 4 mmol g^{-1}). In many cases this will be the method of choice for preparing the most stable and high activity catalysts.[26–30] Catalysts made by this method which are described here include amino-functionalised mesporous silicas which are useful solid bases. Post-modification methods can be used to build the required functionality. Grafting can also result in the formation of a range of surface species resulting from binding *via* one, two or three Si–O–Si groups, attachment of oligomeric silanes and the presence of physisorbed material, which must be thoroughly washed off before the catalyst can be used in reactions. Loading is often at the low end of the range resulting in the need for large quantites of catalyst. Furthermore, the surface is hydrolytically sensitive and leaching of organic structures can occur with vigorous hot aqueous solvent washing as well as in the presence of silicophilic nucleophiles such as F^- and OH^-. Nonetheless, many catalysts have been successfully prepared by the grafting method and it remains the most commonly used method for the preparation of catalysts based on organically modified silicas.

More robust catalysts can be prepared *via* initial surface chlorination followed by reaction of the Si–Cl (or other surface–Cl) groups with organometallic reagents such as Grignards to give directly bonded surface–alkyl and surface–aryl functions.[17] This method gives a more chemically resistant direct surface Si–C bond and avoids the problems of formation of surface-bound oligomers and variable modes of attachment. The method is, however, less frequently used than grafting, largely because the technique is more difficult. Initial surface chlorination is most efficiently carried out in a fluidised bed reactor using an organochlorinating agent (*e.g.* CCl_4) at high temperatures (typically 400–450 °C). More simply but less efficiently, the support can be refluxed with thionyl chloride. The resulting chlorinated silica (or other support) is moisture sensitive and must be handled with care if full advantage is to be taken of the active sites (Si–Cl loadings of > 2 mmol g^{-1} are easily achieved by the fluidised bed reactor method). The subsequent reaction with an organometallic also requires due care and, more seriously, limits the functionalities that can be attached. Again this can be overcome by futher reaction of the surface groups, although this would mean a minimum of three steps being required to achieve the desired surface structure.

4   Properties of Supported Reagents

Methods of Studying Supported Reagents

There are numerous analytical techniques available for the study of supported reagents. Among the more commonly available are Fourier transform infrared spectroscopy (FTIR), especially *via* diffuse reflectance techniques (DRIFTS), titration methods, thermal analysis – notably thermogravimetric analysis (TG) and differential scanning calorimetry (DSC) – or differential thermal analysis (DTA), and surface area analysis and porosity. More specialist techniques include high resolution solid state nuclear magnetic resonance spectrometry, using magic angle spinning techniques (MAS NMR), X-ray diffraction (XRD), scanning and tunnelling electron microscopies (SEM and TEM), and photo-electron spectroscopies (PES). These techniques and the key properties of supported reagents that they can reveal are summarised in Table 4.4.

Surface Structure

The nature of the surface of the solid catalyst is ultimately vital to its performance and value. The number of catalytically active sites, their dispersion over the surface of the solid, their accessibility to substrate molecules and their activity or strength are fundamental properties that can be influenced by the nature of and any pretreatment of the support, the method of preparation of the

Table 4.4 *Techniques for studying supported reagents*

Technique	*Information obtained*
Surface area/porosity	Total surface area; pore volume; pore size distribution; pore shape
Infrared spectroscopy [especially diffuse reflectance IR (DRIFTS)]	Identification of surface species; level of hydration
Ultraviolet–visible spectroscopy (especially DRUVS)	Identification of surface species; measurement of surface polarity *via* dye adsorption
Titration methods (including spectroscopic titration)	Measurement of acid, base, oxidation and reduction sites
Thermal analysis (including evolved gas analysis and temperature programmed desorption)	Thermal stability; dehydration/drying temperature; phase changes; heats of adsorption; identity of adsorbed species
Electron microscopy (SEM and TEM)	Surface morphology; presence of microclusters; crystallinity; presence of channels/pore structure
X-Ray diffraction	Reagent dispersion; bulk structure (regular solids)

catalyst, its activation and, for reuse, the reaction system. The nature of the unmodified surface of the support in a supported reagent can also be very important since it can have an overall influence on surface polarity and hence adsorption characteristics and catalytic turnover rates.

The number of active sites will be directly related to the available surface area of the solid and the efficiency of any surface treatment (*e.g.* in the preparation of a chemically modified support material). The original principle behind the preparation of the first supported reagents was to achieve an effective increase in the surface area of the reagent. This increase will be dependent on:

1. The surface area of the support material.
2. The dispersion of the reagent.
3. Structural changes resulting from any reaction between the support and the reagent.

The surface areas of common support materials and other solids can be readily measured by the Brunauer–Emmett–Teller (BET) method which relies on the adsorption of a monolayer of an inert gas such as argon or nitrogen. Commercial automatic surface area analysers and porosimeters can accurately measure surface areas ranging from $< 1 \text{ m}^2 \text{ g}^{-1}$ to $> 1000 \text{ m}^2 \text{ g}^{-1}$. It is possible to measure surface areas manually using gas burettes but this is a very slow, labour-intensive method.

Making full use of the available surface area on a support material by achieving maximum coverage of active sites requires good dispersion. In the case of physisorbed inorganic reagents, it is practical to aim for dispersion that is good enough to avoid the formation of microcrystals on the support surface. As a general guide, for typical support materials with a surface area of 100–500 $\text{m}^2 \text{ g}^{-1}$, an idealised monolayer of a fairly small reagent would correspond to *ca.* 1 mmol g^{-1}. Excess microcrystalline reagent can be observed by powder XRD and SEM, although neither technique is particularly sensitive (XRD detectability, for example, can be considered good enough to see crystals greater than 50 Å thick). In some cases spectroscopic techniques can be used to distinguish between reagent molecules at the support surface and other not directly adsorbed reagent molecules. Examples of this include supported azides, cyanides and thiocyanates, and metal oxo anions such as MnO_4^-, whose infrared spectra are sensitive to environment and have characteristic bands that occur in different positions dependent on whether they are on the support surface or in a crystal.[31–33] With chemically modified supports prepared by post-modification, it is normal practice to use excess (typically silane) reagent so as to achieve the highest possible loading. With simple silica supports (surface areas of 300–800 $\text{m}^2 \text{ g}^{-1}$) this method rarely gives surface loadings greater than 0.5 mmol g^{-1} even with quite small molecules.[17] Sol-gel methods incorporating a functional group in the formulation (*e.g.* an amino group for subsequent derivatisation[16]) are known to give higher loadings, especially in the preparation of the new hexagonal mesoporous silicas.[17,29] Achieving good dispersion of the catalyst sites is a far from trivial task especially with porous solids that have a heterogeneous internal structure (broad pore size distribution and range of pore structures). Large pores are likely to be filled first, although it is also true that large pores are also likely to be emptied first (*e.g.* on volatilisation of the reagent). It may be possible to take advantage of these facts and prepare a material with reagent molecules concentrated in the smaller pores (which should give better shape selectivity characteristics in reactions). In practice, very slow deposition of the reagent molecules should help achieve good dispersion. The increasing use of the more structured mesoporous solids (as well as zeolitic solids) will also lead to a greater use of better dispersed catalysts.

Structural changes to the support material leading to significant changes in surface area are relatively rare but important in some particular cases. Expandable clays are vulnerable to irreversible collapse of their layered structures on heating to about 300 °C.[14] Pillared clays are generally more robust materials. Other high surface area solids can also lose surface area on heating to higher temperatures. Thus the surface area of zirconia, for example, is reduced from *ca.* 180 m^2 g^{-1} to as little as *ca.* 20 m^2 g^{-1} on heating to 700 °C. The presence of reagents is likely to effect this so that acid-treated (sulfated) zirconia, for example, can maintain a higher surface area on calcination, possibly due to the formation of a more microporous structure.[34] Small, high charge density anions can be corrosive to support materials. Silicas are readily attacked by F$^-$, OH$^-$ and CN$^-$ due to the high affinity of silicon for these anions (strong Si–F, Si–O and Si–CN bonds). Silica-supported fluorides, for example, are mostly ineffective as solid bases or as sources of nucleophilic fluorine.[15,21] Alumina is more robust and KF–alumina, for example, is a widely used solid base (see later), although excess fluoride and high activation temperatures lead to the formation of K$_3$AlF$_6$.[21,35] The surface area of such supports can be dramatically reduced to a small fraction of the original value by the attack of these anions.[35]

Site accessibility is likely to be an important issue in most types of heterogeneous catalysis. Amorphous solids will have a range of site locations with accessibility dependent on the local environment such as the shape and size of the pocket, fault or pore in which individual sites are located. Solids with regular structures may still contain small pores which some molecules may find difficult to access due to size or polarity. Thus microporous zeolites with pore diameters less than 1 nm cannot allow the passage of large molecules. Just as significantly, the diffusion rates of most molecules in the liquid phase will make them largely unsuitable as catalysts for many liquid-phase applications due to poor reaction rates. Use of larger pore materials either in the form of amorphous solids (typical silica gels, for example, have average pore diameters of 2–20 nm) or as structured solids such as hexagonal mesoporous silicas (which can be synthesised with pore diameters of about 1.6–10 nm), can help alleviate these problems. In all cases, however, molecular and surface/in-pore polarities are very important factors. The highly polar nature of most of the catalysts described in this book will result in strongly preferential adsorption by polar molecules. This can lead to slow diffusion of non-polar molecules through porous catalysts, even when the pores are an order of magnitude greater in size than the molecules.[36] In extreme cases, a very polar product may strongly bind to and block active sites, effectively stopping catalytic turnover. An example of this is the failure of powerful solid Lewis acids to catalyse Friedel–Crafts acylation reactions, at least at moderate temperatures.[37] The strong Lewis acid–Lewis base (ketone product) complex remains intact under normal reaction conditions. Even in less extreme cases, it is still quite likely that the reactions will be mass transport limited, with the rate-limiting step being adsorption of non-polar substrate molecules and/or the desorption of polar product molecules. The formation of small polar molecules, notably water, as side products in reactions such as condensations can also be highly influential since such molecules will only slowly disperse from active sites. The same factor means that efficient drying of solid catalysts prior to use is likely to be very important, especially in the early stages of the reaction. Indeed, induction periods have been observed in, for example, the oxidation of hydrocarbons by inefficiently dried mesoporous catalysts containing immobilised metal complexes.[38,39] Effective drying of the catalysts can completely remove the induction period. Presumably it takes a finite time for the water to disperse from the highly polar catalyst pores into the non-polar liquid phase. The future success of porous solid catalysts in many liquid-phase reactions will depend on methods being devised to alleviate these problems. The use of organofunctionalisation of solid surfaces to reduce local polarity[40] is one potentially valuable approach, as is the use of membranes that can selectively enable the removal of, for example, small polar molecules. Thus the partial organofunctionalisation of titanium-doped MCM materials using trimethylsilane has a dramatically positive effect on the ability of the materials to catalyse alkene and alkane oxidations using hydrogen peroxide.[41] The original principle behind the preparation of supported reagents, that of enhancing activity through dispersion of a reagent/catalyst over a high surface area support, gives an oversimplified view of the materials as we understand them today. Indeed, the activity of a supported reagent ASR can rarely be determined as a simple sum of the activities of the component parts (AS + AR). This complexity is, of course, of value when there is a synergistic effect between the support and the reagent so that

$$ASR > AS + AR$$

This synergistic effect can be in addition to that due to a higher effective surface area of the reagent. With physisorbed materials, unexpectedly high activity can result from high local concentrations of reagent molecules in pores or other constrained regions of the support. Thus zinc chloride supported on acid-treated clays is remarkably active in some Friedel–Crafts alkylation reactions: 'clayzic' enables simple benzylation reactions to be completed within minutes at room temperature, whereas both the clay and the zinc chloride individually require several hours to catalyse these reactions.[42,43] This is not believed to be due to the creation of new highly active sites during the preparation of the material. Indeed, spectroscopic titration (see below) of the surface sites shows the presence of only quite weak Lewis acid sites.[36] A low concentration of Brönsted sites is also evident, although this is unlikely to be significant since clayzic is most effective in catalysing alkylations using alkyl halides rather than alkenes, and the analogous silica-supported catalyst is just as effective but has no measurable Brönsted acid sites.[44] Rather it is likely that the synergy is due to high local concentrations of zinc chloride (presumably Zn^{2+} in particular) within the mesopores of the support creating high reaction fields which are highly activating towards incoming alkyl halide molecules.

Many of the more recently developed supported reagent catalysts are chemisorbed in nature with the active sites being chemically bonded to the surface. It is more difficult to compare the activites of these molecules to unsupported analogues. Enhanced activity can still be expected from local in-pore effects resulting from high site concentrations and it is likely that the presence of isolated sites will result in activities different to those of a comparable free reagent or indeed to those of a comparable reagent in solution (where intermolecular associations between reagent molecules may occur). A good general illustration of this is with supported transition metal complexes. The relatively flat surface of the larger pores of support materials should largely preclude a full coordination of the metal centres by chemisorbed groups unless the tethered metal can back-bite to an available site on the surface. This is unlikely to occur with long spacer groups and especially with the more rigid spacer groups such as those containing aromatic functions that have been employed in some of the more recently reported examples.[17,25] This coordinative unsaturation can be satisfied by loosely bound solvent or water molecules which should be easily removed on thermal activation leading to enhanced site activity in subsequent reaction. Interestingly, a synergistic effect has been observed by forcing two active sites close together through the use of a bifunctional surface-bound group; a so-called bicipital supported phase transfer catalyst, with adjacent phosphonium centres, shows unusually high activity in some reactions, presumably due to a cooperative effect between the cationic centres.[45]

There are several possible methods for studying the nature of the surface and adsorbed reagents including FTIR, MAS NMR and DRUV. Diffuse reflectance FTIR is particularly useful for giving valuable information quickly although the amount of information it can provide is material dependent. It is particularly useful in confirming the presence of IR-active adsorbates, including organic functionalities, and hence can also be used to check for the destruction of surface sites due to thermal or chemical damage. Undesired reactions between the support and adsorbed species such as with fluorides supported on alumina (which give Al–F species) can also be easily observed by FTIR. Strong physisorption may also lead to changes in the IR spectrum of the adsorbate such as those resulting from the change in symmetry of polyatomic species.[15]

A particularly powerful application of FTIR is the study of the sites on solid catalysts *via* probe molecules. The technique is often referred to as spectroscopic titration. UV–visible spectroscopy and MAS NMR spectroscopy can also be used to titrate the sites on catalyst surfaces.[15,46,47] Applications include:

1. The study of the nature and strength of the sites on solid acids
2. The study of surface polarity and other 'solid solvent' properties
3. Identifying electron transfer sites

The IR spectroscopic titration of solid acid sites is a particularly powerful technique.[47] Basic probe molecules that are IR active can be used to recognise and titrate different types of acid site. Thus pyridine gives different IR bands depending on whether it is binding to Lewis acid, Brönsted acid or simple hydrogen bonding sites (Table 4.5). Furthermore, the position of the band due to pyridine–Lewis acid complexes is a measure of the strength of that complex and hence the strength of the Lewis acid site. The bands due to the pyridinium ion are fixed in position but their relative intensity can be used to give at least semi-quantitative information on the concentration of the Brönsted acid sites. Other small basic molecules such as ammonia and acetonitrile can also be used to probe solid acid sites.[15]

Table 4.5 *IR absorption band values (cm^{-1}) for pyridine (py) in the spectroscopic titration of acids*

Hydrogen-bonded py (Lewis)	Coordinatively bonded py (free)	Pyridinium (Brönsted)
1440–1447	1447–1460	
1485–1490	1488–1503	1485–1500
1580–1600	1580	
	1600–1633	1620
		1640

Numerous coloured or UV-active molecules can also be used as indicator molecules to probe solid acid surfaces, although they are generally used to give a total acid strength rather than more subtle information about the nature of sites. In many cases, suitable indicators undergo visible colour changes on protonation and they can be studied by eye. Typically a series of indicators is applied until a colour change does not occur, showing the upper limit of the acidity of the catalyst (Table 4.6).

Solvatochromic parameters determined by measuring the effect of a solvent on a range of UV-active indicator molecules are now commonly used to help define solvent 'power' and to help predict or explain solvent properties.[48] Some progress has been made in this context with solids. Reichardt's dye is a single probe molecule that gives a remarkable range of absorption maxima values (453–810 nm) depending on the polarity of the medium in which it interacts. The maximum absorption for the dye has been shown to vary over a wide range of values on adsorption on silicas that have been chemically or thermally treated. Silica gels give high values consistent with very polar surfaces (due to the high concentration of surface hydroxyls). Generally, the dye indicates

Table 4.6 *Some useful indicators for measuring the acidity of solid surfaces*

Indicator	pK_a
2,4-dinitrotoluene	−13.8
3-chloronitrobenzene	−13.2
4-chloronitrobenzene	−12.7
anthraquinone	−8.2
chalcone	−5.6
dicinnamalacetone	−3.0
2-nitrophenylamine	−2.1
4-phenylazodiphenylmethane	+1.5
4-phenylazo-1-naphthylamine	+4.0
neutral red	+6.8

reduced surface polarity on organofunctionalisation of the silica surface. Heating also causes a reduction in apparent surface polarity, presumably due to the condensation of neighbouring hydroxyl groups (Table 4.2).[20]

Temperature programmed desorption (TPD) relies on the interaction between a site on the solid and a probe molecule causing an elevation on the desorption temperature of the probe. Solid acids are commonly studied by this method. Thus the temperature of desorption of ammonia can be correlated with the acidity of a solid:

> 700 K	very strongly acidic
500–700 K	strongly acidic
400–500 K	moderately acidic
< 400 K	weakly acidic

Acidities measured in this way can correlate extremely well with acidities measured by other techniques such as *via* indicators.[49] Weaker bases can be used to probe strong acid sites. Remarkably, benzene may not desorb from the very strongly acidic sites on some sulfated zirconias until over 850 K. Substituted benzenes can also be used and a series of these ranging from the moderately basic 1,3,5-trimethylbenzene (mesitylene) to the very weakly basic fluorobenzene can be used reminiscent of a range of coloured indicators. A solid acid doped with ammonia may well give several desorption maxima representing several sites of different acidities. The technique can therefore be used to measure site homogeneity as well as strengths.

Catalyst Stability

The stability of any catalyst is ultimately of considerable importance, since it will affect its lifetime and its regenerability and it may affect process work-up and product purification. Of immediate concern in many applications, and particularly with catalysts containing immobilised metals, is leaching. Not only can this lead to a competing homogeneous contribution in the reaction, but it will also lead to added impurities in the product mixture and will reduce catalyst lifetime. Product contamination with heavy metals is likely to be totally unacceptable in many commercial processes where metal contaminants may have to be below ppb levels. The total absence of a homogeneous contribution can be very difficult to prove, especially at very low levels. Leaching is likely to be of particular importance when polar products or substrates are being used. It is generally advisable to condition the catalyst fully by thorough washing at close to reaction conditions. The stability of the catalyst under reaction conditions typically of above ambient temperatures can be tested by removing the solid from the hot reaction mixture (*e.g.* by hot filtration) and then testing the filtrate for any residual activity. This 'hot filtration test' makes allowance for catalytically active species that might leach during reaction then readsorb on cooling.[50]

Decomposition, sintering or crystal growth are factors of significance only when high temperatures are employed in the preparation of the catalyst (*e.g.* in the conversion of a supported metal salt to the metal oxide or the metal) or when the adsorbed reagent is unstable.

Catalyst poisoning is a common problem. With porous solids, this can take the form of pore blockage, which effectively prevents access to the majority of active sites, ultimately restricting the chemistry to the external surface. This can take the form of 'coking', although this tends to be less of a problem in liquid phase reactions than in the higher temperature gas phase reactions. The former are, however, more likely to suffer from site poisoning resulting from the adsorption of polar molecules. A solid catalyst may not function because it quickly adsorbs a low-concentration polar impurity from the reaction mixture. Polar adsorbates are a particular problem in reactions involving relatively non-polar substrates which may not be able to displace the poison from the active catalyst sites. The most common example of a polar adsorbate poison is water. It should be appreciated that supported reagents which are based on porous polar hydrophilic materials such as silica and clays will efficiently remove water from most organic environments. Pre-drying of both the catalyst and the substrates are sensible pre-treatments, although there will be cases where an aqueous environment might be beneficial to reaction, for example by helping to wash away polar products from the catalyst. In reactions which produce water as a co-product (some oxidations and base-catalysed reactions, for example), the water should be continuously removed from the reaction system if at all possible (typically by Dean–Starck traps). It may be possible to remove organic poisons by desorption through washing or heating but this would normally have to be carried out in a separate step, although it need not require removal of the catalyst in the case of fixed-bed type reactors. Of course, thermal clean up must take place below the decomposition temperature of the catalyst itself. Fortunately, many supported reagent-type catalysts, and even those with organic functions built on their surface, are stable to quite high temperatures (650 K or higher).

Catalyst Recovery and Regenerability

It is commonly stated that solid catalysts are more easily separated from liquid reaction mixtures than homogeneous catalysts. While this is essentially correct, the situation can be easily oversimplified by the assumption that separate liquid and solid phases can be efficiently separated by a simple procedure such as filtration. Particulate solids may be easily recovered in the laboratory by filtration (although care must be taken with very small particles which can go through conventional filters). On a commercial scale, however, filtration may not be a practical option. Among the simplest options is decantation, although this relies on effective (and preferably quick) settling of the catalyst particles and the catalyst will be left wet with reaction liquors. This can be a suitable method if the vessel is to be recharged with fresh substrate mixture (and solvent where required). A popular option is to use non-particulate forms of the catalysts; the catalyst could be present in the form of a membrane, for example. Catalytic membranes are very interesting materials since they not only present the catalyst in a convenient way, avoiding any need for a solid–liquid separation step (other than washing), but they can also be used to separate two liquid phases such as aqueous and organic phases commonly used in phase-transfer catalysed reactions.[51] One interesting example of this is the use of capsule membrane PTC using nylon capsules. The capsules are impregnated with the catalyst, effectively heterogenising it. They are then filled with the organic phase and subsequently put in the aqueous phase. Reaction occurs at the surface of the capsule facilitated by the immobilised catalyst.[52] Catalytic plate reactors are also available and again remove the need for a separation stage. Of course, the most common solution to the problem is to go away from batch mode and operate the process continuously through a catalyst bed in a so-called 'plug flow' type reactor (see Chapter 1). This is the technology of choice in gas-phase processes, but is rarely used in liquid-phase reactions. Rapid throughput of the liquid is essential and may be difficult to achieve. Nonetheless, we must expect a growing use of this type of technology whereby the catalyst may stay in the reactor for long periods of time (> 1 year is not uncommon in some gas-phase processes) before it needs to be removed. More sophisticated reactors are available and suitable for liquid-phase reactions. The spinning disc reactor,[53] for example, could be adapted to work with a solid catalyst (impregnated on the disc), especially for rapid reactions.

The regenerability of the catalyst refers to its reactivation and reuse. Solid catalysts based on porous solids will lose their activity on application due to a number of possible factors, including poisoning, pore blockage or loss of active sites through thermal degradation, chemical destruction or leaching. At some point in a continuous reaction or at the end of a batch reaction, a decision has to be made on the fate of the catalyst. If it is economically viable or environmentally preferable, the catalyst can be recovered for reuse. Normally, some loss in activity or selectivity will have occurred and a reactivation stage will be required. Most simply, this will take the form of washing (to remove any 'difficult' organics) or heating (typically air calcination to volatilise off loosely bound organic poisons or burn off more resistant organics). Both procedures, though simple, can only be applied if the catalyst itself is sufficiently robust to these treatments. Supported reagents based on physisorption only cannot be washed with solvents in which the reagent has any significant solubility. Organically modified support-based catalysts will probably only tolerate moderate temperatures for thermal reactivation. If the loss in activity is due to decomposition or through leaching into solution, then the minimum treatment required will be a repeat of the preparation process so as to reintroduce more active sites. In continuous processes based on a fixed bed of catalyst, regeneration is technically easier than if the catalyst needs to be recovered (*e.g.* by filtration); the flow of substrate(s) is interrupted and replaced by a flow of washing solvent or by a heating stage, the catalyst being regenerated *in situ*.

5 Applications of Supported Reagents

There are numerous applications for supported reagent catalysts in organic synthesis.[3,4,6,7,15,21,22,43] These will be illustrated in sections on the important areas of partial oxidations, acid-catalysed reactions and base-catalysed reactions with other areas illustrated in a final miscellaneous section.

Partial Oxidations

Catalytic oxidants based on porous inorganic and other solids should be distinguished from stoichiometric supported reagents. The latter have been well studied but are unlikely to have any significant value beyond the laboratory scale.[4,6,15] Catalytic supported reagents can be divided into three categories: supported metallic species, metal ion-exchanged materials (discussed in Chapter 2 and in the case of pillared clays, in Chapter 3) and those based on organically modified solids.

Remarkably, quite subtle differences in the preparation of a metallic species can determine if the resulting material acts catalytically or stoichiometrically in oxidation reactions. Thus while most reported forms of chromium complexes supported on inorganic and organic solids are stoichiometric oxidants, a catalytically active form can be prepared by addition of $K_2Cr_2O_7$ to alumina under carefully controlled conditions of pH and temperature.[54] The resulting material contains a low (<0.1 mmol g^{-1}) level of chemisorbed chromium which is resistant to washing with water although it is removed on treatment with aqueous acid making it unsuitable for use in many oxidations with aqueous peroxide, for example. The exact nature of the immobilised chromium is unknown. The chemisorbed material can be used to catalyse the aerial oxidation of diphenylmethanes to benzophenones and of ethylbenzenes to acetophenones.[55] Reactions are typically performed in the absence of solvent and with air as the only consumable source of oxygen, so that when the organic reactions are entirely selective, the only waste product will be water. In practice, the organic reactions are indeed highly selective with little side-product (typically $<5\%$) formed from over-oxidation, for example. The catalyst is now a

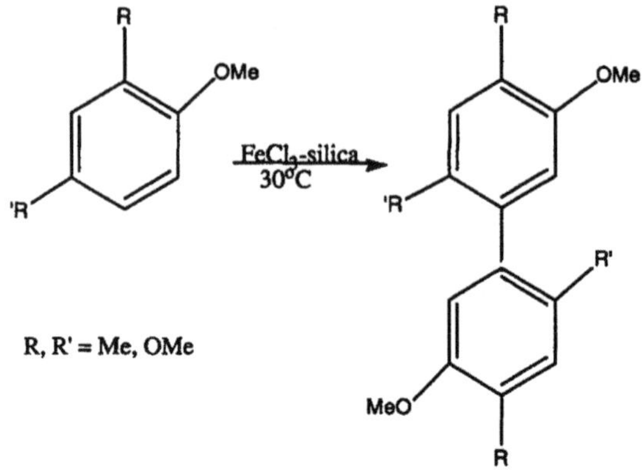

Figure 4.1 *Supported iron(III) chloride-catalysed coupling of aromatics*

commercial product and has been shown to run with little loss in activity, in semi-continuous alkylaromatic oxidations.[55]

Supported $FeCl_3$ will catalyse the oxidation of activated phenols and the coupling of aromatics such as that shown in Figure 4.1.[56] It is interesting to note that $FeCl_3$ is inactive in this reaction. Similarly $FeCl_3$–alumina has been successfully used to synthesise several symmmetrical and unsymmetrical hexa-alkoxytriphenylenes (very important as discotic liquid crystals) from dialkoxy-benzenes and terphenyl derivatives.[57]

Copper(II) sulfate supported on alumina has been sucessfully used to catalyse the oxidation of benzoins under microwave.[58] Microwave radiation can drastically reduce the reaction time and enable milder conditions to be used in a wide range of reactions catalysed by solids. Full industrial exploitation is, however, yet to be realised.

Highly dispersed titanium oxide species on silica prepared by the sol-gel method catalyse the selective epoxidation of propene by molecular oxygen.[59] This is potentially very significant as the new commercial route to propene oxide is based on the reaction of propene with hydrogen peroxide catalysed by a mixed Ti–Si oxide; the direct reaction with oxygen has clear advantages.

Active manganese dioxide supported on alumina has been reported as showing dramatic improvements in ease of handling while maintaining the activity and selectivity of the reagent in alcohol oxidations.[60] It is not clear however, if the supported reagent is catalytically active or simply a stoichiometric reagent.

Other catalytically useful supported metallic species include molybdic acid on charcoal, which is quite useful for epoxidations,[61] and SeO_2 on silica, which will catalyse the oxidation of allylic methyl groups with organic hydroperoxides as the source of oxygen.[62]

Oxidation is the area where organically modified surfaces have probably found most use. A wide variety of materials have been prepared, using several methods for attaching the organics to the surface, with many areas of oxidation chemistry benefiting from the materials thus derived.

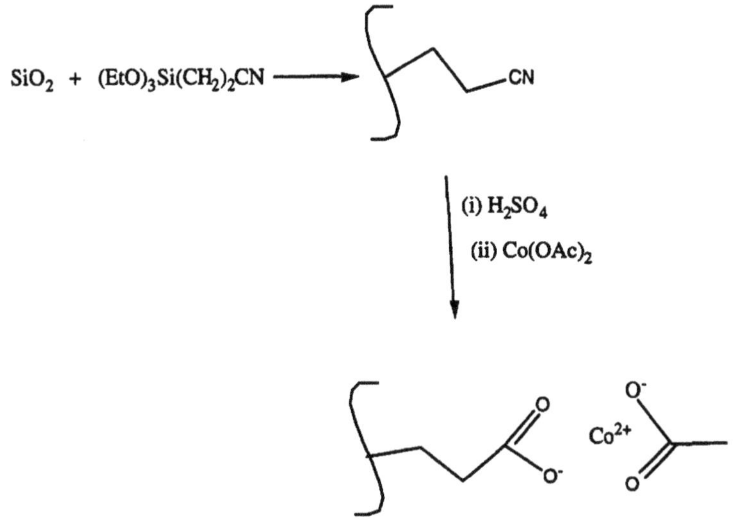

Figure 4.2 *Preparation of supported cobalt acetate*

A simple immobilised form of Co(OAc)$_2$ has been prepared which is capable of the epoxidation of alkenes.[63] This material was prepared by reacting an amorphous silica with a cyano-functionalised trialkoxysilane followed by post-modification of the material so as to convert the CN group to a CO$_2$H group, which can then be used to bind Co^{2+} ions *via* ion exchange (Figure 4.2).

This is a particularly useful result, since the organofunctionalised silica is shown to survive the harsh conditions required to hydrolyse the nitrile group (50% H$_2$SO$_4$ at reflux, 24 h). The material was found to catalyse efficiently the epoxidation of alkenes using sacrificial aldehydes and oxygen (Table 4.7).

A related catalyst type where the metal centre is supported by a longer spacer chain has been reported (Figure 4.3).[64,65] This material was prepared from aminopropyl-silica by reaction with terephthalaldehyde to form the mono-imine (attachment of the amine to the surface precludes reaction at both ends of

Table 4.7 *Epoxidation of alkenes using supported cobalt acetate*[a]

Alkene	Time (h)	Yield of epoxide (%)[b]
cyclohexene	5	85
oct-1-ene	5	45
octa-1,7-diene	24	48[c]
2,4,4-trimethylpentene	5	95
hex-1-ene	24	30
styrene	3	32[d]

[a] All reactions carried out at 19 °C in dichloromethane with isobutyraldehyde as sacrificial aldehyde. [b] GC yield with internal standard. [c] Monoepoxide; 7% of diepoxide was also formed. [d] 5% PhCHO and 21% polymer also formed.

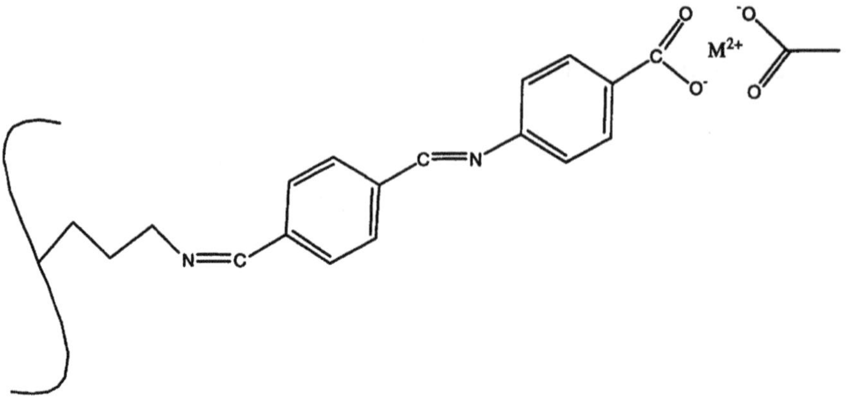

Figure 4.3 *Supported metal complexes based on silicas chemically modified with a double imine*

the dialdehyde), followed by formation of a second imine with *p*-aminobenzoic acid. The supported acid was then treated with metal acetates to generate the active catalyst.

The catalyst is active in the same epoxidation reaction as the supported Co acetate above. Interestingly, the Ni version is most active, followed closely by Cr and Cu, with Mn and Co being significantly less active. Of particular interest is the ability of this catalyst to carry out the oxidation of alkyl aromatics. In this case, the Cr version of the catalyst is the best, and allows a conversion rate of 370 turnovers h^{-1} to be achieved.

Silica chemically modified with salicylimine will effectively complex metal ions. The heterogeneous metal complexes show some activity in the oxidation of cyclohexane and in alkylaromatics.[66]

Silica gel functionalised with *N,N*-dimethyl-3-aminopropyltrimethoxysilane and complexed with metal ions including Fe(III), Co(II) and Ni(II) is effective in the room temperature oxidation of the C–H bond in cyclohexane. The system involves molecular oxygen, zinc and acetic acid. Product yields based on cyclohexane of up to 1.4% have been observed.[67]

Active epoxidation catalysts based on supported metal complexes with the triazacyclononane ligand system have also been reported.[68] Both silicas and MCMs were used as supports. The catalysts were prepared by the reaction of the cyclic ligand with a supported glycidyl material in the case of the MCM, and with both glycidyl and chloropropyl in the case of the silica. The ligand system was then modified by reacting the two free amines with propylene oxide, followed by metal complexation. The epoxidation of styrene was used as a test reaction. Selectivities and turnover numbers (mol h^{-1}) were higher in the case of the MCM-derived materials than the silicas, regardless of the nature of the supported ligand.

The use of MCM-type supports is likely to increase. Higher loadings are commonly observed compared to amorphous silicas and greater thermal stabilities can result from incorporating the functional group (which is subsequently converted into the active group) into the sol-gel preparation. This was nicely illustrated with solid peracids prepared *via* materials containing surface cyano functions.[69] While these are not strictly catalytic, it is worth noting that while a loading of <1 mmol g^{-1} is achieved with amorphous silica, loadings of *ca.* 3 mmol g^{-1} can be obtained from the sol-gel approach [starting with a mixture of Si(OEt)$_4$ and (EtO)$_3$Si (CH$_2$)$_3$CN].

The grafting of a titanocene on to a MCM-41 support has been used to prepare a powerful epoxidation catalyst for substrates including pinene and smaller unsaturated molecules (Figure 4.4).[70]

This builds on the key research carried out on titanium-doped mesoporous solids and their use in selective oxidations using hydrogen peroxide. Catechol and hydroquinone are now manufactured commercially using the Enichem catalyst TS-1 (prepared from tetraethylorthosilicate and tetraethylorthotitanate[71]) and hydrogen peroxide.[72] Traditional methods of manufacture have low atom efficiency and the successful exploitation of new greener catalytic methods in such processes is very important.[73,74]

HMS-supported Mn(III)(Cl)Salpr exhibits reasonable activity as a catalyst in the epoxidation of styrene using PhIO as the oxidant in MeCN. No reaction occurs when using hydrogen peroxide or sodium hypochloite as the oxidants. Unfortunately there is relatively low selectivity to the epoxide due to the formation of by-products mostly adsorbed on the catalyst surface.[25]

A particularly active catalyst for the aerial oxidation of alkylaromatics is

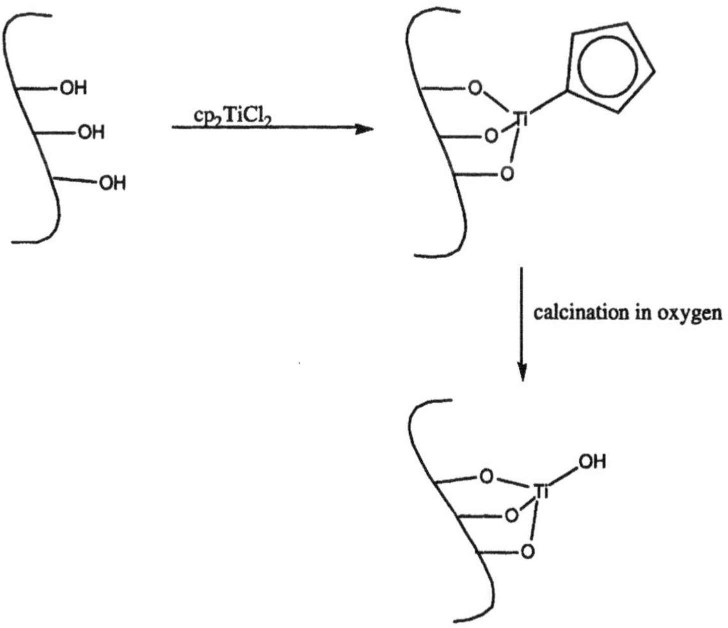

Figure 4.4 *Preparation of MCM41-grafted titanocene-derived Ti catalyst*

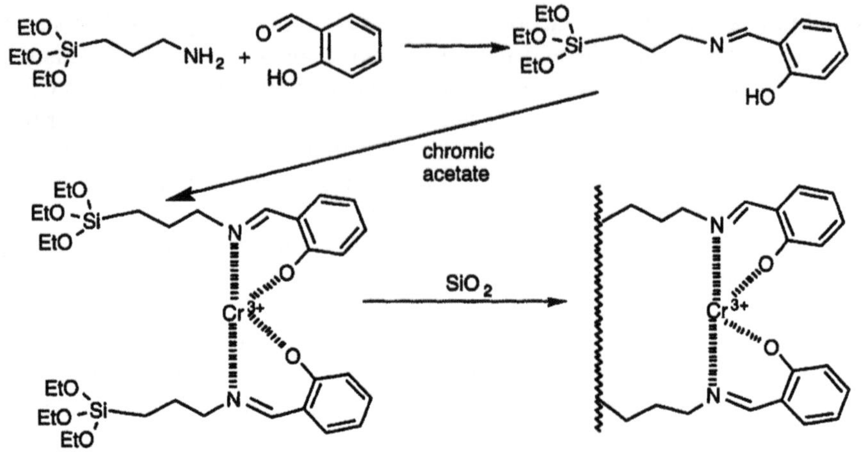

Figure 4.5 *Preparation of Schiff base-supported chromium catalyst*

based on an immmobilised Schiff base complex.[75] The catalyst is prepared in a somewhat different way to most other chemically modified mesoporous solids. The Schiff base complex of a chromium(III) salt possessing pendant triethoxy-silane functions is first prepared in solution and then reacted with a silica surface (Figure 4.5). In this way the complex is effectively imprinted onto the surface. The resulting loading of catalytic sites is low (*ca.* 0.1 mmol g^{-1}) but the sites are strongly bonded and resistant to leaching even in acetic acid. Higher loaded materials can be prepared using sol-gel methods. The catalyst can be used to promote the selective side chain oxidation of various alkylaromatics including the commercially important oxidation of *p*-xylene to toluic acid and teraphthalic acid (Table 4.8). Reaction occurs at a moderate rate even at amospheric pressure and the only by-product is water. Analogous physisorbed catalysts are significantly less active, revealing the effectiveness of the imprinting method for making robust and active catalysts.

As indicated above, metalloporphyrins have also been the subject of much

Table 4.8 *Oxidation of alkylaromatics using the imprinted Schiff base chromium catalyst*[a]

Substrate	Temperature (°C)	Product(s)	Isolated product yield (%)
ethylbenzene	130	acetophenone	50
p-xylene	145	*p*-toluic acid	29
		terephthalic acid	5
o-xylene	145	*o*-toluic acid	7
p-chlorotoluene	130	*p*-chlorobenzoic acid	12

[a] Reactions carried out with 1.5 g catalyst/4.1 mol neat substrate under high shear with a fast air feed for 24 h.

work, and several routes have been developed to attach them to heterogeneous supports.[76–79] Porphyrins are expensive, and thus recovery becomes economically important. It is also possible that attachment to a surface may hinder destructive oxidation of the electron-rich ring system, a factor which traditionally limits their useful lifetime. The use of charged groups, typically ammonium or sulfonate attached to the periphery, has been used to enhance adsorption to polar supports such as silica and magnesium oxide. Direct covalent binding to silica surfaces has been achieved by coordinative binding of the metal centre to supported imidazoles, pyridines, *etc*. The second mode of attachment is *via* aryl groups attached to the periphery of the ring system. Some examples of immobilised porphyrins are shown in Figure 4.6.

Another approach is the nucleophilic displacement of chloride from chloropropyl-silica with a pyridine-substituted porphyrin. These materials are active in the epoxidation of alkenes, where iodosylbenzene is the preferred oxidant, and in the oxidation of alkanes to alcohols and ketones. The copolymerisation of a porphyrin containing four attached trimethoxysilane groups with tetraethoxysilane, leading to an active hybrid silica-porphyrin, offers another route to these important catalysts.

Recently, controlled procedures for the covalent anchoring of iron tetrasulfophthalocyanine onto amino-modified silicas have been developed to fix the complex either in a monomer or dimer form.[80] Remarkably, the normally inactive dimeric oxo iron tetrasulfophthalocyanine is an active and selective

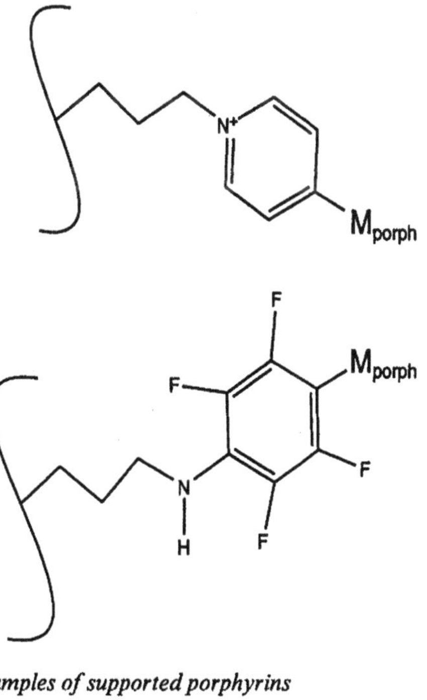

Figure 4.6 *Some examples of supported porphyrins*

Table 4.9 *Catalytic oxidation of 2-methylnaphthalene by* t-*butyl hydroperoxide*

Catalyst	Conversion in 24 h (%)	Selectivity to vitamin K_3 (%)
homogeneous FePcS	64	11
homogeneous FePcS	61	10
m-FePcS-silica	58	11
d-FePcS-MCM41	81	24
d-FePcS-silica	78	24

catalyst for the oxidation of 2-methylnaphthalene to 2-methylnaphthaquinone (vitamin K_3) using *t*-butyl hydroperoxide, when it is supported onto aminpropylsilica (Table 4.9). The catalyst can also be used in the oxidation of 2,3,6-trimethylphenol to trimethylbenzoquinone.

One problem relating to the oxidation of hydrocarbons with hydrogen peroxide is the difficulty of having appreciable concentrations of the non-polar substrate and the polar oxidant together at the catalytic centre. One approach is based on the attachment of a mixture of poly(ethylene oxide) and poly(propylene oxide) to a silica, followed by the physisorption of methylrhenium trioxide (Figure 4.7) allow the efficient mixing of both reaction partners.[81] The material catalyses the efficient epoxidation of alkenes with hydrogen peroxide.

A mix of the two polymeric chains was shown to be better than a single type of chain. This is attributed to a combination of the relatively hydrophilic poly(ethylene oxide) and the more hydrophobic poly(propylene oxide) giving the right balance of properties, and allowing optimum mixing of the two reagents. We should also not completely lose sight of the many significant articles reporting the use of polymer-bound metals in selective oxidation reactions including palladium cross-linked polystyrene for mild toluene oxidations,[82] vanadium(V) and chromium(VI) immobilised on gycidyl methacrylate (GMA-EGDM) for various oxidations including benzylamine to benzaldehyde,[83] and polymer-tethered manganese (salen) complexes for chiral epoxidation.[84]

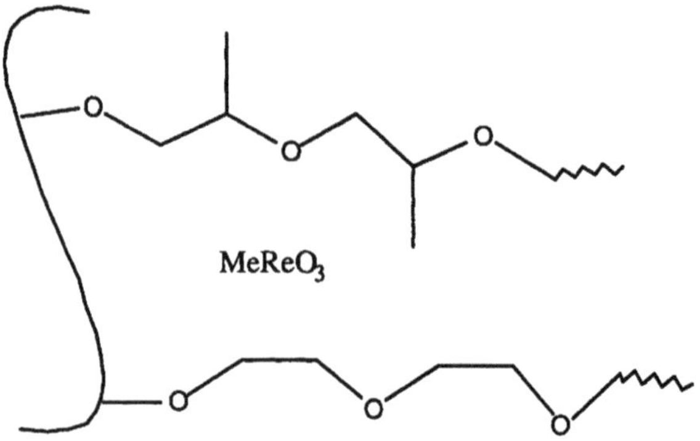

$MeReO_3$

Figure 4.7 *Methylrhenium oxide supported on a polyether-modified silica*

Finally it should be noted that caution is required in interpreting the data from supported metal-catalysed oxidation reactions. It has been shown on several occasions that catalyst instability under oxidation conditions can lead to metal leaching. The homogeneous metal may then contribute or even dominate the catalysis. One example of this is the oxidation of alcohols with hydrogen peroxide–urea complex using a zeolite-supported vanadium picolinate catalyst which the authors have recognised as probably being due to leached vanadium. Other proven examples of leached metal-catalysed oxidation reactions exist. It should be noted that even when the catalyst is shown to be apparently stable to leaching with the reaction system components, during the oxidation a change in the oxidation state of the metal can significantly affect its stability in the heterogeneous phase.[50,85]

Reactions Catalysed by Solid Acid Supported Reagents

The use of solid acids has been traditionally biased towards large-scale continuous vapour phase processes such as catalytic cracking and paraffin isomerisations. However, it is increasingly recognised that there is also a great need for solid acid catalysts which are effective in liquid-phase organic reactions such as those employed in many batch-type reactors by fine, speciality and pharmaceutical intermediate chemical manufacturers. This has contributed towards a substantial recent research effort into the development of new solid acid catalysts.[86–91]

Some of the major reaction types which are important in this context are Friedel–Crafts alkylations, acylations and sulphonylations, aromatic halogenations and nitrations, and isomerisations and oligomerisations. These reactions are currently catalysed by mineral acids such as H_2SO_4 and HF and by Lewis acids such as $AlCl_3$ and BF_3. These reagents are hazardous in handling, damaging to plant through their corrosiveness and add process difficulties through the use of quenching and separation stages which lead to large volumes of toxic and/or corrosive waste. Solid acids based on organic polymers such as ion-exchange resins have some important applications but rather poor stability. More robust polymers such as Nafion show promise but their cost will preclude them from widespread application. Organic–inorganic composite type materials can partly alleviate this, however.[92]

Inorganic supported reagents such as 'clayzic' (acid-treated clay supported zinc chloride)[42,93–95] have a limited but valuable range of applications, notably in Friedel–Crafts benzylations (Figure 4.8), where a remarkable synergism between the individually weakly active components results in a very active catalyst.

Clayzic also shows a number of interesting and quite suprising reactivity trends:

- the order of reaction rates for the benzylation of halobenzenes using undried clayzic (PhI > PhBr > PhCl > PhF > PhH) is the opposite to that using homogeneous $AlCl_3$;[96]

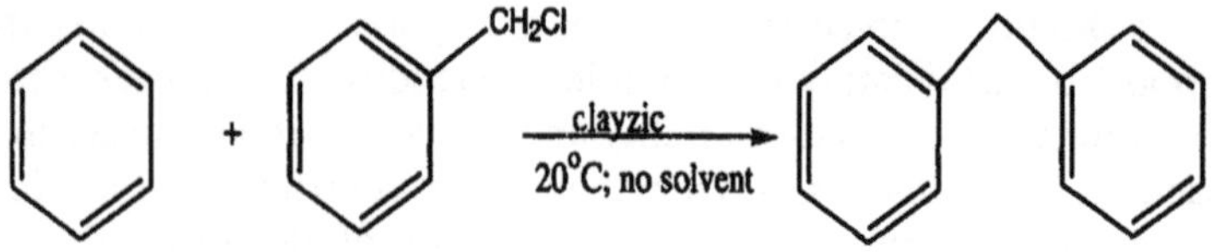

Figure 4.8 *Benzylations using clayzic*

- whereas $PhCH_2Cl$ reacts more quickly than $PhCH_2OH$ in individual reactions, the alcohol is entirely consumed first in a mixed reaction system;[97]
- the benzylation of anisole using benzyl chloride gives a non-linear Arrhenius plot.[98]

The first two of these observations can be explained by the highly polar clayzic pores into which the more polar or polarisable substrate will diffuse. The third of these observations is consistent with the presence of two catalyst sites, Lewis and Brönsted. At low temperatures, the former are complexed by anisole and do not take part in the catalysis. Indeed, spectrosopic titration of the acid sites on clayzic using pyridine confirms the presence of the two types of site although the Brönsted acid activity is very low. Interestingly the strength of the Lewis acid sites appear to be low and the high activity of clayzic in benzylation reactions is best explained by very high in-pore concentrations of reagent molecules giving high local reaction fields rather than high activity of individual sites.

Clayzic is also capable of catalysing a range of other alkylation reactions (Table 4.10).[7,21] Generally the mesoporous silica version of clayzic has a higher optimum loading for zinc chloride and is more active.

Unfortunately, clayzic and its analogues are not particularly active in other Friedel–Crafts reactions. The catalysts require forcing conditions (> 100 °C and long reaction periods) to promote acylations (including benzoylations) and sulfonylations. Reasonable product yields based on quite low quantities of catalyst can be obtained, however, and combined with the non-toxic nature of the catalysts, these offer not insignificant advantages over conventional Friedel–Crafts catalyst such as aluminium chloride. Supported zinc chloride is the basis of a commercial Friedel–Crafts catalyst.[99]

Other applications for clayzic include the preparation of benzothiophenes by cyclisations of phenylthioacetals (normal catalysts can cause extensive poly-

Table 4.10 *Alkylation reactions catalysed by clayzic*

Alkylating agent	Aromatic substrate	Comments
$PhCH_2Cl$	ArH	Fast reactions
Ph_2CHCl	ArH	Slow at room temp
CH_2Cl_2	ArH	Reasonable rates at 40 °C
$(CH_2O)_2$	ArH	Reasonable rates at 80 °C

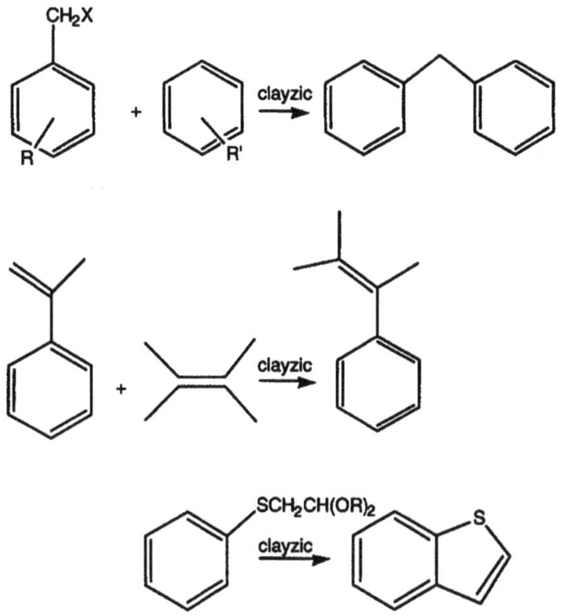

Figure 4.9 *Other reactions catalysed by clayzic*

merisation of the thiophenes and the pores in clayzic are believed to favour the desired intramolecular cyclisation at the expense of the polymerisation[100]) and the olefination of benzaldehyde (involving a previously unknown reaction mechanism).[101] These reactions are summarised in Figure 4.9.

Supported zinc bromide has also been found to have value as a catalyst in the selective bromination of aromatic substrates using bromine.[102,103] Other mild Lewis acids supported on porous inorganic supports, notably silicas and acid-treated clays, have been used as supported reagent catalysts. Commercial catalysts in this category are now available for acetylations, benzoylations and sulfonylations and other reactions, although their activities are somewhat low.[104–109] The development of active and truly catalytic, heterogeneous alternatives to traditional soluble or liquid acids is a very important goal in green chemistry. Materials based on chemically modified inorganic solids have emerged in recent years and show considerable promise. Aluminium chloride is one of the most widely used inorganic reagents in organic chemistry. It is highly soluble and very active. However, its many drawbacks, such as its corrosive nature, difficulties in separation and recovery, and the large volumes of environmentally hazardous waste associated with its use, make it a prime target for heterogenisation.[37] Active heterogeneous forms of aluminium chloride have been reported over several years.[110–116] The most effective of these for liquid-phase applications is believed to contain a mixture of OAlCl$_2$ and O$_2$AlCl sites on the surface. This is prepared by reaction of a surface-hydroxylated high surface area mesoporous inorganic solid such as silica gel or acid-treated montmorillonite with aluminium chloride in a low-polarity aprotic organic solvent.

The optimum loading for a high surface area mesoporous silica is about 1.5 mmol g^{-1}, twice as high as that for the acid-treated montmorillonite clay K10. The former catalyst is also a little more active and selective towards monoalkylation, although K10 is a less expensive support material.

Catalysts prepared in the vapour phase or from a CCl$_4$ solution have proved to be active in gas-phase reactions, but their activity in liquid-phase reactions has generally been poor.

Various mesoporous materials such as MCM-41, MCM-48, SBA-1 and KIT-1 have been rendered more acidic by treatment with reagents including ethanolic solutions of AlCl$_3$ and Al(NO$_3$)$_3$ and slurries of Al(OPr^i)$_3$ in non-polar solvent (*e.g.* hexane) followed by calcination of the resulting solid at temperatures of > 800 K to give solid acids.[117] These treatments create either framework or non-framework aluminium centres which can act as Lewis acid catalytic sites. The materials are more commonly associated with vapour-phase reactions such as cracking rather than liquid-phase organic reactions.[118]

The activity of the catalyst shown in Figure 4.10 in the alkylation of benzene with alkenes is comparable to that of aluminium chloride, but it shows improved selectivity towards monoalkylation compared to AlCl$_3$ itself and is readily recoverable and reusable (unlike AlCl$_3$, which needs to be removed from the reaction after one use, typically by a water quench). The alkylation of alkylbenzenes, and to a lesser extent halobenzenes, can also be carried out using supported aluminum chloride (Table 4.11).

More recently, it has been shown that it is possible to extend the methodology for supported aluminium chloride for liquid-phase applications to hexagonal mesoporous silicas as supports (HMSs and including MCM materials).[115,116] Activity in the alkene alkylation of alkylaromatics is again comparable to that of AlCl$_3$ itself, and the solid acids are also easily recovered and can be reused. Most significantly, the increase in selectivity towards monoalkylation through

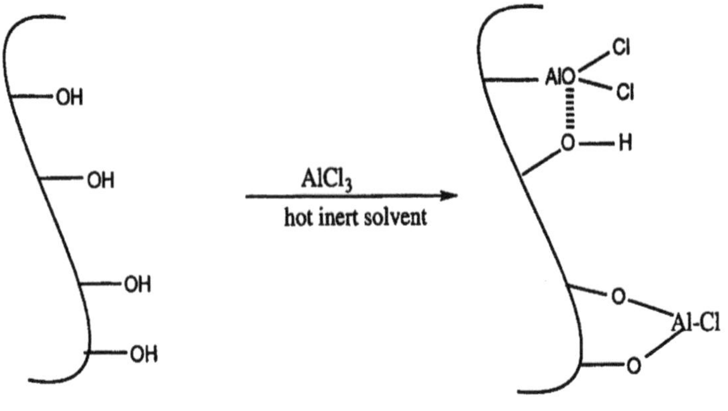

Figure 4.10 *Preparation of supported aluminium chloride for liquid-phase organic reactions*

Table 4.11 *Activity of (optimised) supported aluminium chloride solid acids in the alkylation of aromatic substrates*[a]

Substrates	Catalyst	Reaction time (h)	Monoalkylaromatic (% by GC)
PhH + oct-1-ene	AlCl₃	2	61.6
PhH + oct-1-ene	SiO₂(70 A)-AlCl₃	1.25	78.3
PhH + oct-1-ene	K10-AlCl₃	2	76.3
PhH + hex-1-ene	K10-AlCl₃	2	69.2
PhH + dodec-1-ene	K10-AlCl₃	2	77.3
PhH + dodec-1-ene	K10-AlCl₃	2	71.0
PhMe + oct-1-ene	K10-AlCl₃	1.5	80.9
PhEt + oct-1-ene	K10-AlCl₃	3.5	74.3
PhF + oct-1-ene	K10-AlCl₃	4	29.6
PhCl + oct-1-ene	K10-AlCl₃	4.5	14.2

[a] Reactions carried out at 20 °C with 2:1 molar ratio of ArH to alkene and 10 g catalyst/mol ArH.

the use of the heterogenised Lewis acid is further enhanced (Figure 4.11). By using external site poisons such as Ph₃N (to block external acid sites through complexation) or Ph₃SiCl (to destroy external hydroxyl groups), there is a still greater increase in selectivity, with close to 100% monoalkylation being achievable with the larger alkenes. The extension of the phenomenon of shape selectivity from the small molecules (transformed by zeolites) to large molecules is clearly very important. It promises significant improvements in product selectivity, while maintaining the relatively fast reaction rates that mesoporous catalysts can give in liquid-phase organic reactions.

Boron trifluoride is also widely used as a Lewis acid in industrial processes. It is less active than aluminium trichloride but has the advantages of being more tolerant of air exposure (it is deactivated but not destroyed by water). It is normally used as a complex, such as the etherate, which gets around the difficulties of working with a gas. Several versions of supported BF₃ have been reported. The most recent and possibly most interesting for liquid-phase organic reactions involves the complexation of the Lewis acid with the surface of a silica support in the presence of a protic molecule, typically an alcohol (Figure 4.12).[119,120]

The resulting material behaves like a solid Brönsted acid and is active in the alkylation of phenol with alkenes. It is interesting to note that on heating the material loses the complexed alcohol and the resulting material behaves more like a rather weak Lewis acid. The acidity of the catalyst is dependent on the Brönsted base as well as the solvent used in the preparation, as can be seen from the spectroscopic titration of the acid sites using pyridine (Figure 4.13). The catalysts can also be studied by evolved gas analysis using infrared spectroscopy to monitor the off gases as the material is heated. In this way the evolution of the

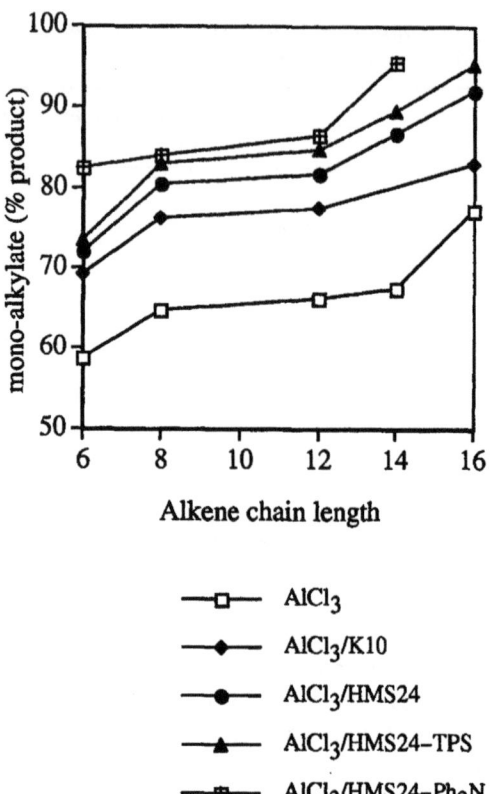

Figure 4.11 *Selectivity* vs *chain length for a series of alkenes with HMS catalysts*

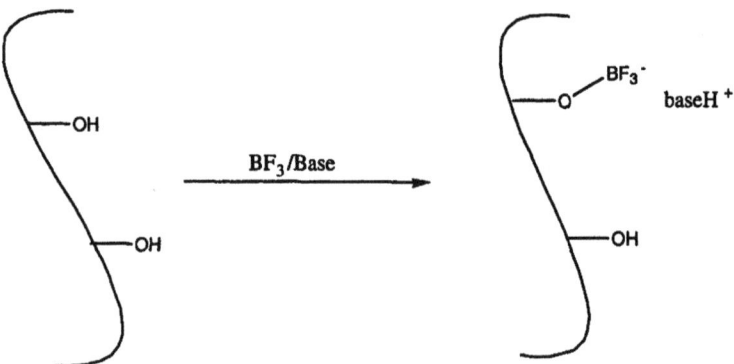

Figure 4.12 *Preparation of a supported boron trifluoride complex*

Brönsted base (converting the catalyst to a solid Lewis acid) and then HF (due to decomposition of the boron trifluoride) can be monitored and used to determine catalyst stability as well as fine tune properties.

More recently, supported boron trifluoride catalysts have been shown to be

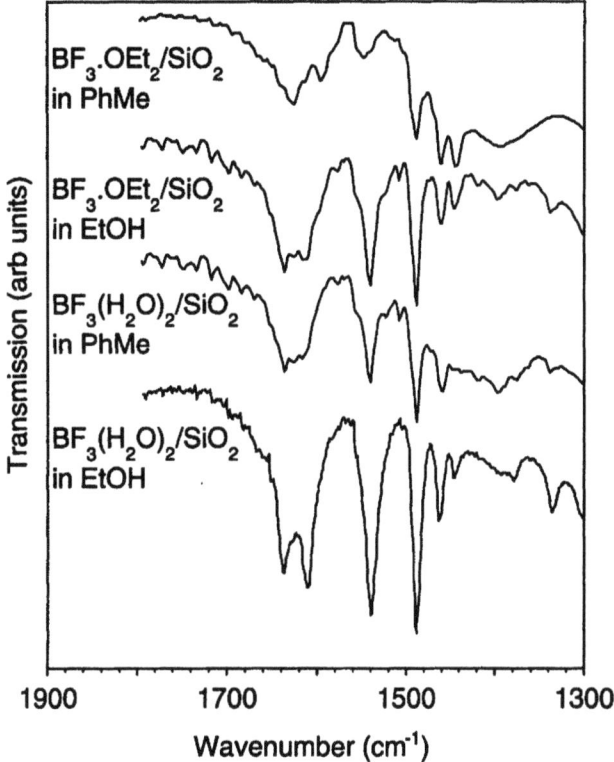

Figure 4.13 *Dependence of the concentration of Brönsted acid sites (as measured by the intensity of the band at 1540 cm^{-1}) on the catalyst preparation for supported boron trifluoride*

active in other liquid-phase organic reactions, including polymerisations, etherifications, esterifications and Claisen Schmidt condensations (Figure 4.14 and 4.15).

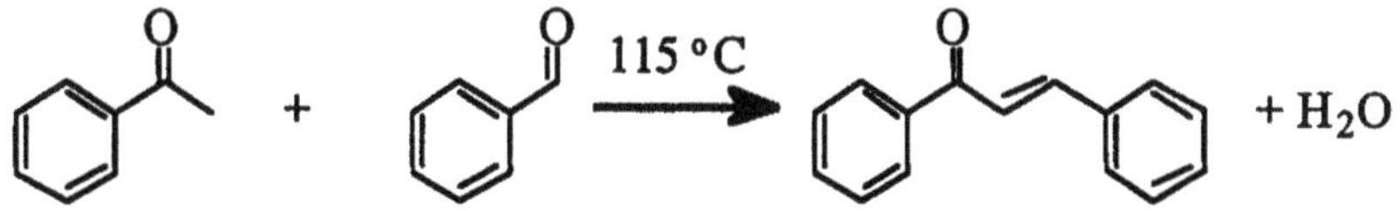

Figure 4.14 *Claisen Schmidt reaction catalysed by supported boron trifluoride*

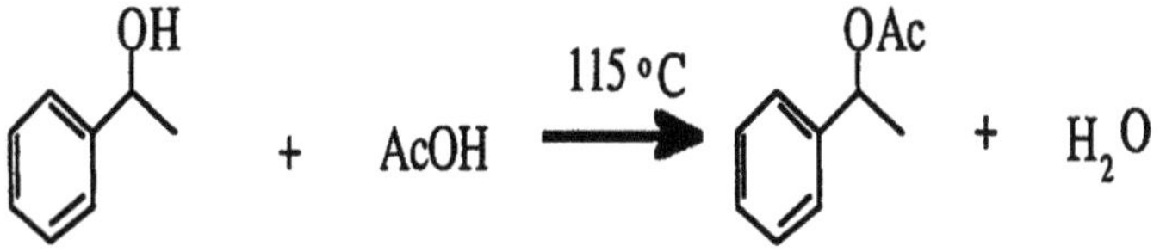

Figure 4.15 *Acetylation catalysed by supported boron trifluoride*

There are a few other references in the scientific literature (as well as a number in the patent literature) to chemically fixed supported Lewis acids. Iron(III) chloride should be reactive enough to form surface $OFeCl_2$ bonds, for example, and a stable form of supported $FeCl_3$ has been reported.[121] The solid acid has been used to catalyse liquid phase Friedel–Crafts benzoylations, although the surface structure and activity on reuse of the catalyst have not been described. At least one commercial form of supported iron(III) chloride is available.[99] Supported SbF_5 has also been extensively studied, although mostly for gas-phase reactions such as the skeletal isomerisation of alkanes where it exhibits high activity indicative of its strong acidity ($H_o = ca. -14$).[122]

Brönsted acids can also be fixed to the surfaces of many common support materials. Surface attached perfluorosulfonic acids have been reported[123] as well as the perfluorinated sulfonic acid resin (Nafion)-silica composites reported a year earlier.[124] The former are prepared by reacting

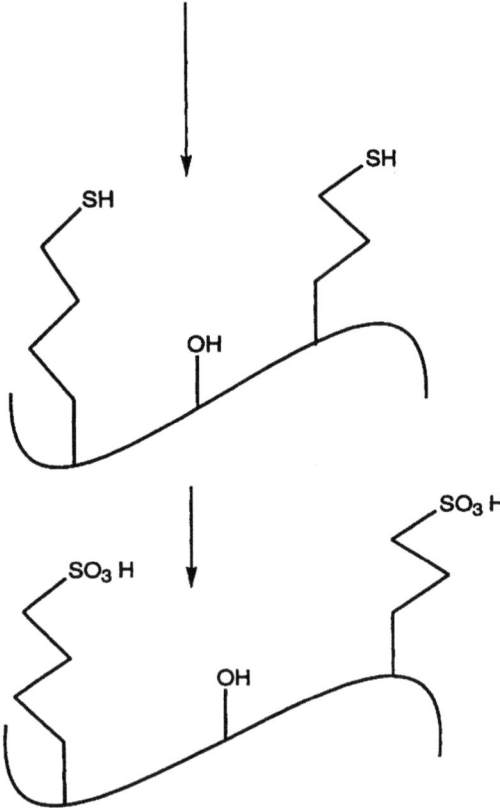

$$(MeO)_3 Si (CH_2)_3 SH + (EtO)_4 Si$$

Figure 4.16   *Preparation of solid sulfonic acid by sol-gel synthesis*

$[(OH)_3Si(CH_2)_3(CF_2)_2O(CF_2)_2SO_3^-]M^+$ with a support material such as an amorphous silica gel.

Alternatively, the solid acid can be prepared by incorporating the same reagent in a sol-gel preparation. The materials are catalytically active for a number of typically acid-catalysed liquid-phase organic reactions. These include the benzoylation of activated aromatics such as *m*-xylene and the alkylation of aromatics using alkenes. They are orders of magnitude more active than conventional acidic ion-exchange resins, but less active than some forms of supported aluminum chloride.[125]

A non-fluorinated solid sulfonic acid has also been reported and shown to be a useful replacement Brönsted acid to sulfuric acid (Figure 4.16).[126]

The solid acid can catalyse reactions for which liquid sulfuric acid is active, including some individually important reactions such as the formation of bisfurylalkanes (Figure 4.17), polyolesters and some generic reactions such as esterifications (Figure 4.18).[127]

The direct reaction of Brönsted acids with support materials has also been used to prepare solid acid catalysts. Simple treatment of silica gel with sulfuric acid followed by mild drying gives a solid acid that is active in various aromatic nitrations (nitric acid or isopropyl nitrate as nitrating agent).[128] Activity of the solid acid is comparable to the more expensive Nafion-H.

Heteropolyacids can also be bound to support materials by direct deposition. More firmly bonded materials can be prepared through chemical surface modification. Acids such as 12-tungstophosphate (PW12) will react with silica gels which have been treated with aminoalkylsilanes.[129] The acidic site of PW12 reacts with the chemically modified surface forming ionic bonds. The latter

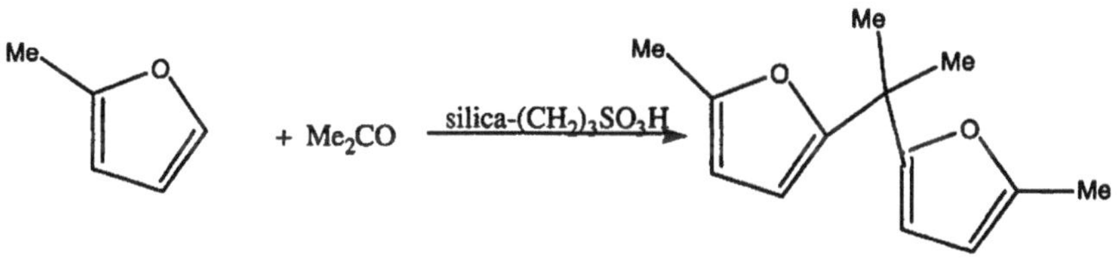

Figure 4.17 *Formation of bisfurylalkanes using solid sulfonic acid*

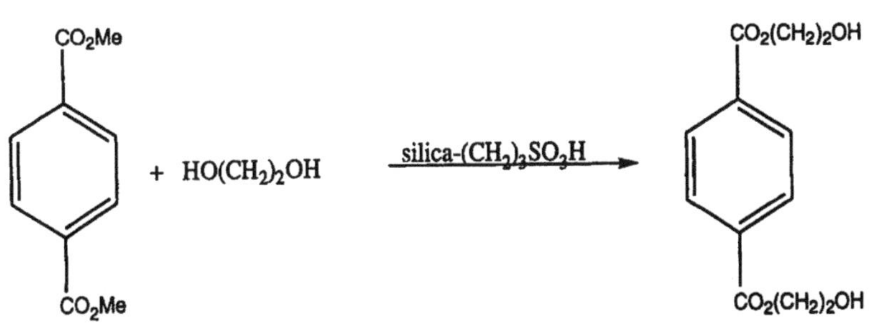

Figure 4.18 *Transesterification using solid sulfonic acid*

materials are generally more active than than supported reagents prepared by direct deposition.[130,131] Further improvements may be achievable by using the structured hexagonal mesoporous materials.

Base Catalysis

Several different supports have been used for preparing supported reagent solid bases, with alumina-based catalysts being the most widely studied in organic synthesis.[6,15,17,21,22,43] Most common basic reagents have been supported including the alkali metals (*e.g.* alumina–Na and silica–Na), alkali metal hydroxides (*e.g.* alumina–KOH) and metal alkoxides (*e.g.* xonotlite–KO-*t*-Bu). Among the more interesting of these are the immobilised alkali metals, which can be prepared in a variety of ways including treatment of the support with a solution of the metal in liquid ammonia. The solids are often brightly coloured, which is due to the formation of colour centres of one electron donor character:

$$Na + [2+] \qquad \rightarrow Na^+ + [+]$$
$$Na + O^- \qquad \rightarrow Na^+ + O^{2-}$$
$$Na + OH \text{ (or } 2OH) \rightarrow ONa + H_2 \text{ (or } H_2O)$$

These materials have been referred to as solid superbases with an estimated Hammett basicity constant, H^-, of >37. They are capable of promoting reactions of hydrocarbons such as the isomerisation of 5-vinylbicyclo[2.2.1]-hept-2-ene to 5-ethylidenebicyclo[2.2.1]hept-2-ene, which is used as a co-monomer in the production of synthetic rubber.[132] Solid superbases have been used in commercial scale manufacturing, where they are typically generated *in situ* so as to overcome their great sensitivity towards the atmosphere.

Attempts to develop supported ionic fluorides and for nucleophilic fluorinations have been largely unsuccessful but have led to the discovery of remarkably useful solid bases with some recent applications showing high regioselectivity.[21,133,134] Metal fluorides such as KF are known to be weak bases and their dispersion over a support leads to a remarkable enhancement in their basicity. The surface chemistry between the adsorbed salt and the support material is quite complex and oxide and hydroxide sites are likely to be formed and to contribute to the basic properties.

When at their most active, supported fluorides are capable of adsorbing large quantities of carbon dioxide from the atmosphere. This and the presence of variable quantities of water which will moderate but not destroy surface basicity (it will irreversibly destroy powerful sites such as oxide but only reversibly hydrogen-bond to fluoride sites) lead to basic properties that are highly dependent on handling as well as preparation. Supported fluorides have been reported as exhibiting very high basicity, but have also been described as weakly basic! Supported fluorides have been used in a wide range of typically base-catalysed organic reactions (Table 4.12) as well as several reactions which require stoichiometric quantities of the base (*e.g.* alkylations).[15,21] KF–alumina is certainly the most widely studied, although important variables such as

Table 4.12 *Reactions catalysed by supported fluorides*

Reaction	Example
Oxidation of alkylaromatics	$Ph_2CH_2 \rightarrow PhCO$
Alkylations	$PhOH + MeOH \rightarrow PhOMe$
Condensations	$EtCHO + MeNO_2 \rightarrow EtCHOHCH_2NO_2$
Rearrangements	$ArCH_2CH{=}CH_2 \rightarrow ArCH{=}CHMe$
Michael reactions	$AcCH{=}CH_2 + EtNO_2 \rightarrow Ac(CH_2)_2CHMeNO_2$
Additions	$CHCl_3 + m\text{-}O_2NC_6H_4CHO \rightarrow m\text{-}O_2NC_6H_4CH(OH)CCl_3$

loading and supported reagent post-treatment remain contentious issues. The basic catalyst is especially effective in carbon–carbon bond-forming Michael reactions.

Some of these reactions have been successfully carried out under continuous flow conditions employing fixed catalyst beds.

Supported reagent bases including KF–alumina can also be used to give regioselective alkylations including the selective benzylation of a pyrazole derivative (Figure 4.19).[133]

Enantioselective alkylations have been achieved using modified MCM-41 materials.[135] Mesoporous templated MCM-41s with covalently linked chiral ephedrine are active heterogeneous chiral auxiliaries in the enantioslective alkylation of benzaldehyde by diethylzinc. Lower rates, selectivities and enantioselectivities are obtained under heterogeneous conditions compared to homogeneous catalysis. This can be explained either by the participation of the uncovered surface to the racemic alkyl transfer or by a restricted accessibility to the catalytic sites in the heterogeneous reactions.

The majority of work on organically modified solids as solid bases has been carried out on the simple 3-aminopropyl-derivatised silica. This material is established as an efficient catalyst for the Knoevenagel reaction.[136–138] There are several interesting features about this application. One of these is the solvent, a

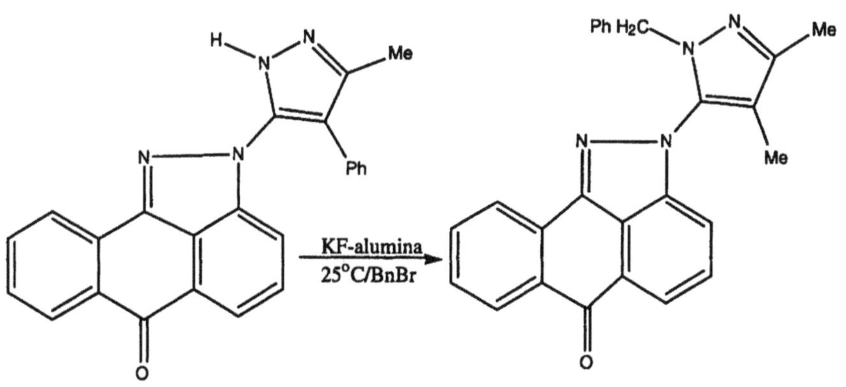

Figure 4.19 *Regioselective benzylation of pyrazole using a supported fluoride*

recurrent theme in heterogeneous liquid-phase catalysis. In the Knoevenagel reaction, the solvent must play two roles. Firstly it must remove water from the system. The reaction is reversible, and the rate of reaction is very likely to be reduced by the presence of water. Secondly, the partitioning of reactants between the solvent phase and the catalyst surface is important and the use of non-polar solvents will allow the reagent to adsorb preferentially onto the catalyst surface. Hydrocarbon solvents such as alkanes can be very effective. The rates of reaction vary with solvent according to the following order:

cyclohexane > toluene > 1,2-dichloroethane > chlorobenzene

This order is valid when the reaction is performed at the boiling point of each solvent. Even the higher temperature used with toluene and chlorobenzene did not bring about a rate close to that of cyclohexane (Figure 4.20).

For the more polar HMS-based solid bases, toluene is the optimum solvent, with partition onto the catalyst surface still being favourable even with the more

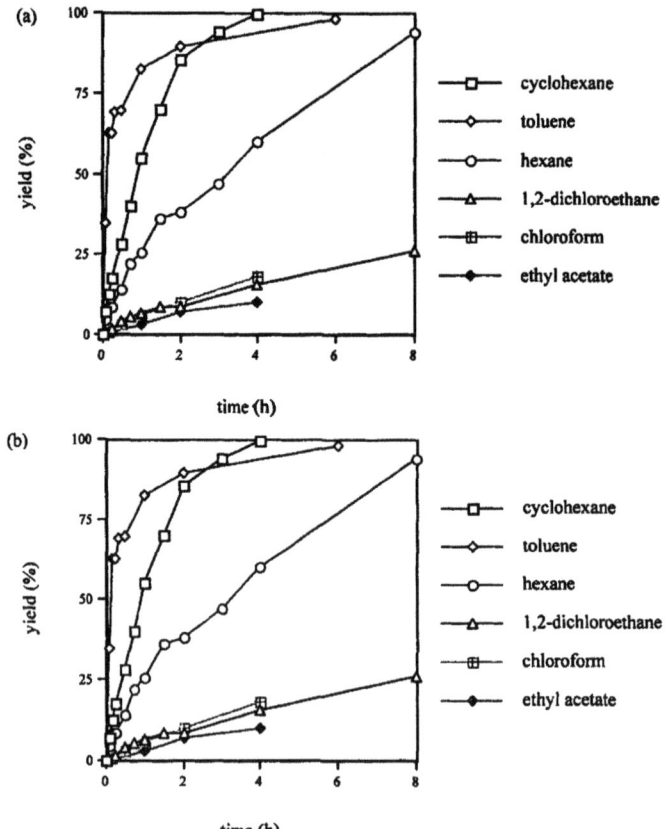

Figure 4.20   *Solvent effects in the Knoevenagel reaction using* (a) *aminopropyl silica and* (b) *aminopropyl HMS catalysts*

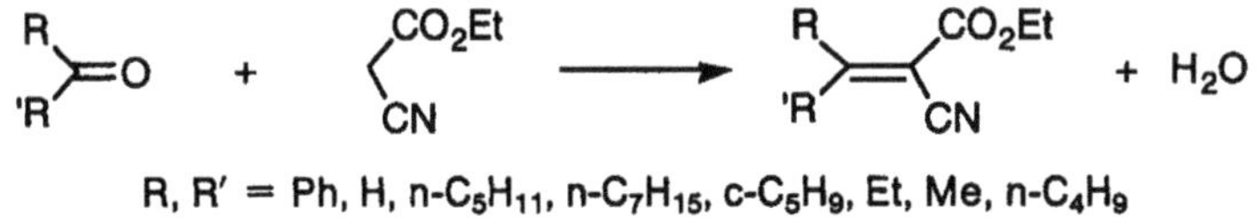

$$R, R' = Ph, H, n\text{-}C_5H_{11}, n\text{-}C_7H_{15}, c\text{-}C_5H_9, Et, Me, n\text{-}C_4H_9$$

Figure 4.21 *Some reactions catalysed by solid bases*

polar solvent. However, increasing the solvent polarity further again reduces the rate dramatically. Various reactions can be catalysed using these solid bases with yields rarely being below 90% and catalyst turnover numbers generally being several thousand (Figure 4.21).[139]

When catalyst poisoning occurs, it is due to amide formation by reaction of the amine with the ester group of ethyl cyanoacetate. The HMS catalysts are generally slightly less active when compared directly, but their ability to function well in toluene, and the much higher loadings achievable (2.5 mmol g^{-1} *vs.* 1.0 mmol g^{-1}) means that under optimum conditions they can match the silica catalysts in terms of rate. Their turnover numbers are typically higher by a factor of 4–5.

Aminopropyl-grafted MCMs have also been used as catalysts with DMSO as the solvent in the reaction of benzaldehyde with ethyl cyanoacetate. A supported piperidine catalyst prepared from reaction of piperidine with chloropropyl-MCM is significantly less active in this reaction, although the different conditions employed in the reactions preclude meaningful comparisons.[140] Studies on the activities of a range of supported amines in the synthesis of monoglycerides from glycidol (Figure 4.22) have shown that solid bases can gain activity on reuse. This has been explained in terms of partial substrate polymerisation on the catalyst surface blocking silanol groups which on subsequent use are less likely to cause side-reactions. In keeping with this theory, it has also been found that blocking underivatised silanol groups by reaction with hexamethyldisilazane improves reaction selectivites to the desired monoglycerides.[141]

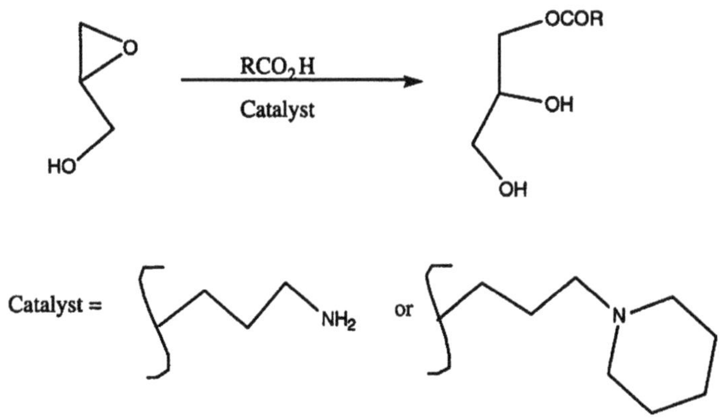

Figure 4.22 *Solid base-catalysed formation of monoglycerides*

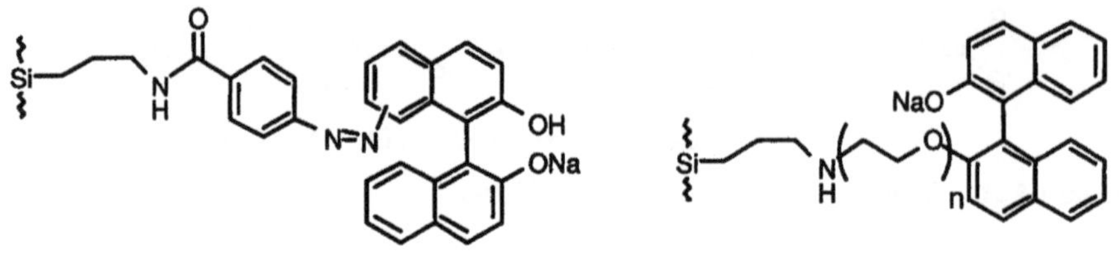

Figure 4.23 *Supported phenolates derived from diazotisation of surface bound aniline and from alkylation of surface bound amine*

Other recently reported supported reagent solid bases include supported phenolates[142] (Figure 4.23) and MCM-41–quaternary tetraalkylammonium hydroxide composites prepared by reacting 3-trimethylpropyl(trimethyl)-ammonium chloride with the mesoporous support. The chloride anions in the latter were then exchanged with hydroxide by treatment with Me₄NOH in methanol. The resulting solid has been shown to be active in Michael additions and aldol condensations and is reported to have good stability.[143]

Other Applications for Supported Reagent Catalysts

While phase-transfer catalysis (PTC) is a well established method with diverse applications in organic synthesis, conventional catalysts suffer several drawbacks including hygroscopicity, low thermal stability and difficulty in separation and recovery. Ironically, the high solubilities of conventional catalysts are a drawback to recovery and a problem to product purification. The concept of triphase catalysis, whereby the catalyst is immobilised onto a support material and the resulting supported PTC is then used in a biphasic aqueous–organic solvent reaction mixture is recognised as a viable solution to many of these problems.[144–146]

Polymer-supported catalysts, especially those based on polystyrene resins, have been used on many occasions but they also suffer from low thermal stability as well as high cost and a tendency to swell in solvent.[147,148] Simple physisorbed supported PTCs can be prepared by the incipient wetting method. In this way, alumina-supported phosphonium compounds have been prepared and used to catalyse various halogen exchange reactions in the gas phase along with various other nucleophilic substitutions (Figure 4.24).

More robust chemisorbed supported PTCs can be prepared in several different ways. Grafting of a silane on to a surface followed by chemical surface modification is one approach (Figure 4.25).[145,149,150] Surface chlorination followed by multi-step chemical surface modification can also be employed (Figure 4.26).[151,152]

Such catalysts are often reusable and are effective in non-polar solvents such as hydrocarbons. This is true even when the unsupported salt is inactive due to poor solubility (*e.g.* Ph₄PBr); the supported catalysts are believed to operate at the aqueous-organic interface.

$$RX \longrightarrow RY$$

$$(X,Y = Cl, Br, I)$$

$$ArO^- + RBr \longrightarrow Ar\text{-}OR + Br^-$$

$$(Ph_3P)_3RhCl + NaOCOR \longrightarrow (Ph_3P)_3RhOCOR + NaCl$$

Figure 4.24 *Some reactions catalysed by supported phase-transfer catalysts*

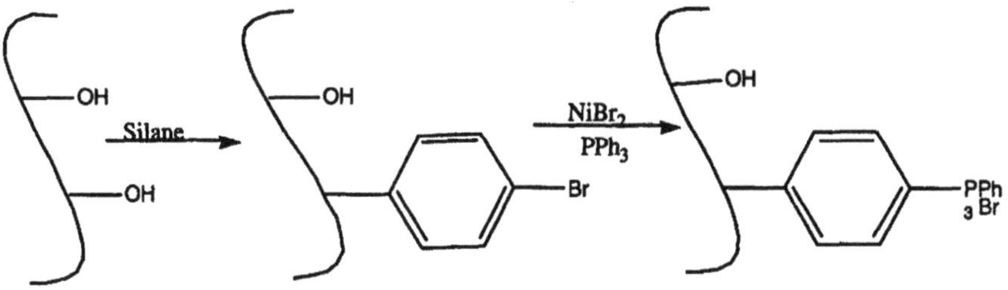

Figure 4.25 *Chemisorbed PTCs derived from initial surface silylation*

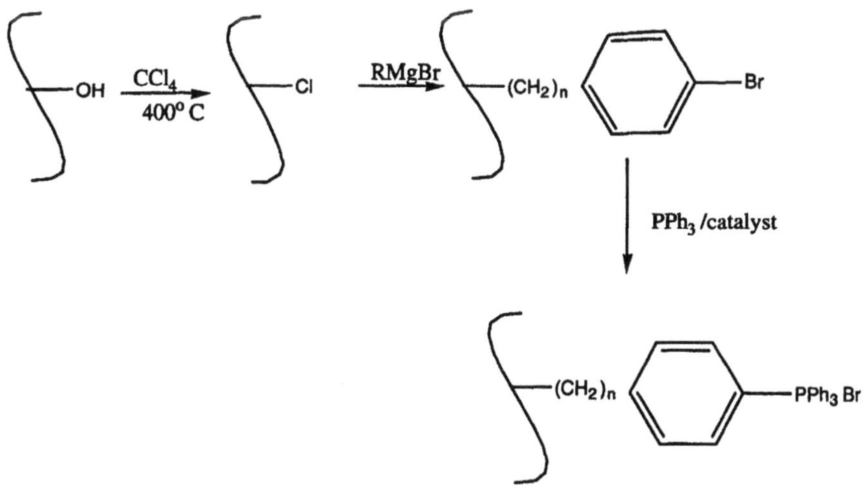

Figure 4.26 *Chemisorbed PTCs derived via surface chlorination*

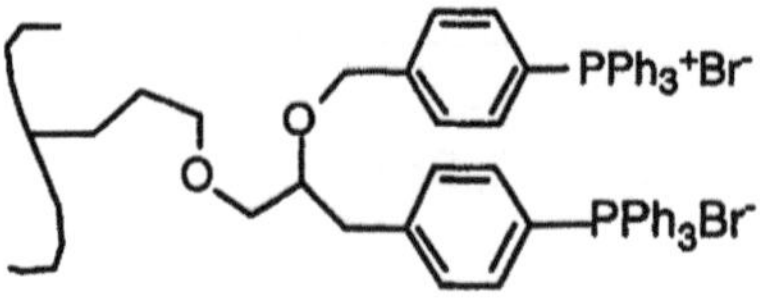

Figure 4.27 *Bicipital-supported phosphonium phase-transfer catalyst*

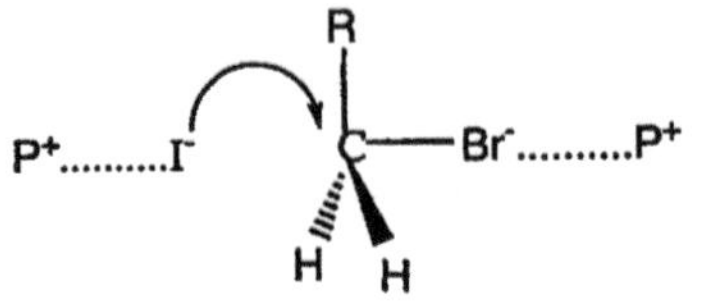

Figure 4.28 *Possible mechanism associated with synergistic effect of bicipital catalyst*

Perhaps the most remarkable of these catalysts is the mesoporous silica-based material which contains two adjacent phosphonium centres referred to as a 'bicipital-supported phosphonium phase-transfer catalyst' (Figure 4.27).[151]

This catalyst is significantly more active than other supported phosphonium salts in nucleophilic halogen exchange reactions. The analogous material with only one of the aromatic rings substituted with a phosphonium group is significantly less active per phosphonium centre than the bicipital material. It is likely that the neighbouring centres can produce a synergistic effect through simultaneous polarisation of the C–Br bond by one phosphonium centre and attack by the I^- delivered by the other cation (Figure 4.28).[153]

This is supported by the unusually low activation enthalpy and entropy found for reactions catalysed by the bicipital material, which are consistent with a lower energy pathway and a more ordered transition state.

The activitities of silica-supported phosphonium are support pore size dependent with *ca.* 100 Å typically giving the most active catalysts. This is very similar to more simple physisorbed silica-based supported reagents and seems to support the view that for liquid-phase reactions catalysed by porous solids, a reasonably large pore is required to give a good molecular diffusion rate.

Chemically modified solid supports can also make very effective ligands for metal ions enabling heterogenisation of many valuable catalyst structures. The immobilisation of palladium and applications of the resulting solid catalysts in important reactions such as hydrogenation, carbonylation, amidation and carbon–carbon forming coupling reactions are good examples of this.[154–159] Apart from organic polymer-based materials, catalytically active clay- and silica-supported palladium complexes are known. Mercaptopropylsilica, for example, can be used to bind palladium chloride, which on reduction with hydrazine gives a highly active and stereoselective catalyst for the arylation of styrene and acrylic acid giving high yields of various *trans*-stilbenes and substituted *trans*-cinnamic acids (Figure 4.29).[155]

A similar supported palladium catalyst is active in the amidation of aryl

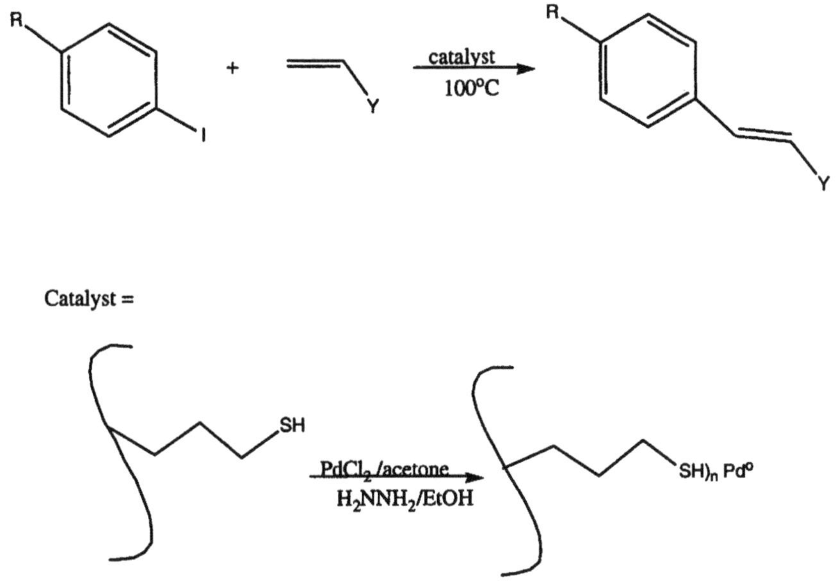

Figure 4.29 *Supported palladium-catalysed arylations*

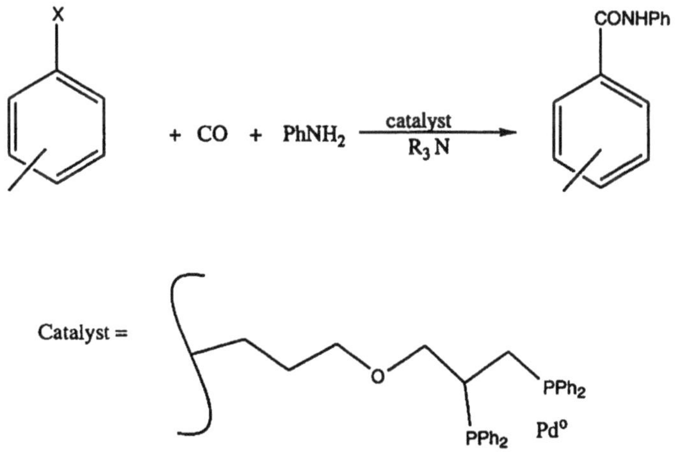

Figure 4.30 *Amidation of aryl halides using a supported palladium catalyst*

halides with carbon monoxide and aniline at atmospheric pressure (Figure 4.30).[156]

Similarly, chemically modified silica-supported palladium catalysts are useful in the phenylation of acid chlorides using sodium tetraphenylborate.[158]

Montmorillonite-based palladium catalysts are active in the hydrogenation of styrene to ethylbenzene under mild conditions (25 °C, 1 atmosphere H_2).

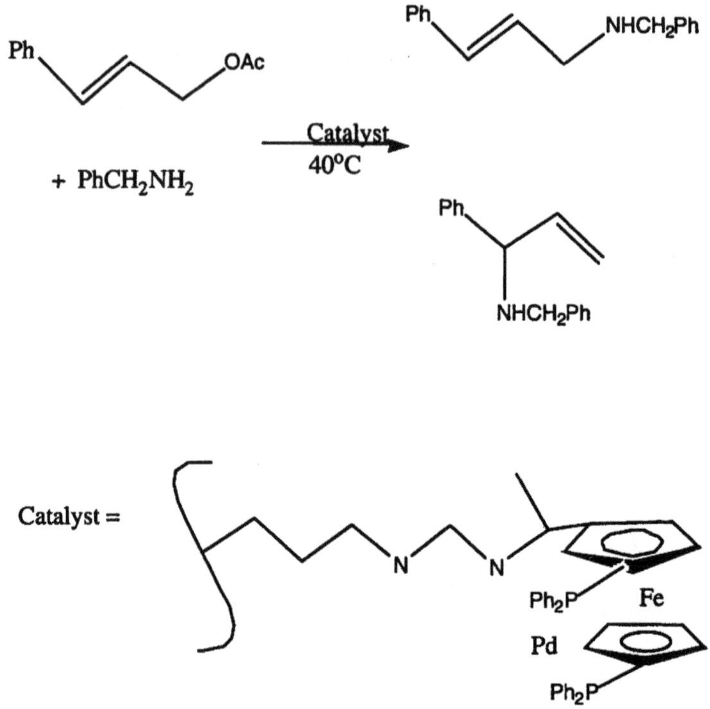

Figure 4.31 *Chiral supported palladium catalyst and its use in enantioselective aminations*

Organic polymer-based catalysts can also be prepared but are generally inferior to those based on inorganic materials.

A chiral palladium catalyst can be prepared by anchoring the 1,1'-bis(diphenylphosphino)ferrocene ligand to MCM-41 and coordinating Pd(II). This catalyst shows a degree of regioselectivity and enantiomeric excess in the allylic amination of cinnamyl acetate (Figure 4.31). This is far superior to that of its homogeneous counterpart or to that of a surface-bound analogue based on a non-porous silica.[159]

Other metals can be similarly immobilised. Silica- and clay-supported rhodium complexes, for example, are effective hydrogenation catalysts.[160,161] An interesting variant on this involves a heteropolyacid to assist the metal-support binding. The heteropolyacid, such as phosphotungstic acid, is attached to the support (*e.g.* montmorillonite) by the incipient wetting technique. The solid material is then treated with a solution of the homogeneous catalyst such as Rh(DiPamp).[162]

An early example of a mesoporous-supported rhodium catalyst is that based on controlled pore glass in which there is a shapely defined pore size of *ca.* 24 Å.[163] Impregnation of such a glass with HRu(CO)(TPPTS)$_3$ (TPPTS = triphenylphosphine trisulfonate) gives a catalyst with activity in hydroformylation reactions:

$$CH{=}CH + CO + H_2 \rightarrow \text{-}CH(CHO)CH_2\text{-} + \text{-}CH_2CH(CHO)\text{-}$$

This is the basis of the very important oxo process used in the manufacture of various carbonyl compounds.

Silica-supported rhodium hydrides are highly efficient isomerisation catalysts. They are prepared by the reaction of silica suspended in toluene with tris(allyl)rhodium. The intermediate complex reacts with hydrogen to eliminate propene and propane to give a material believed to have neighbouring rhodium sites connected by hydrogen bridges.[164]

Remarkably, the heterogenisation process can confer dramatically enhanced activity on the metal complex. Thus, the dimolybdenum complex $[Mo_2(MeCN)_8][BF_4]$–SiO_2 prepared by direct reaction of silica with the metal complex is unusual in its ability to catalyse the polymerisation of norbornene in the absence of an aluminium co-catalyst and at moderate temperatures.[165] Sol-gel processing of the ruthenium complex *cis*-Cl(H)Ru(CO)(P)₃ (where P is a coordinated ether-phosphine) with tetraethoxysilane and [Al(O-i-Pr)₃] gives a stable material with moderate surface area which is active in the hydrogenation of *trans*-crotylaldehyde with reasonable chemoselectivity to the carbonyl reduced products, *cis*- and *trans*-crotyl alcohol.[166]

Enantioselective heterogeneous catalysis is a particularly important goal. The asymmetric aminohydroxylation and dihydroxylation of alkenes has been achieved using a silica gel-supported bis-cinchona alkaloid (Figure 4.32).[167,168] The heterogeneous material has a much higher binding affinity for OsO_4 than the homogeneous analogue and no leaching from the heterogeneous silica gel supported (QN)₂PHAL osmium complex has been measured. The catalyst is completely reusable without any measurable change in enantioselectivity.

Non-metallic chemically modified solids have also been developed for liquid-phase catalytic applications. Silica-supported guanidinium chloride, for example, has been shown to have high efficiency in the decomposition of methyl chloroformate (into CH_3Cl and CO_2) and electrophilic reactions of carboxylic acids and epoxides.[169]

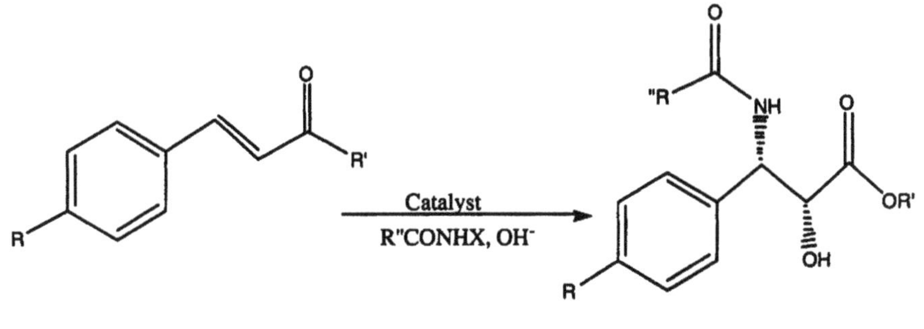

Catalyst = silica-supported bis-cinchona alkaloid/ $K_2OsO_2(OH)_4$

Figure 4.32 *Silica gel-supported bis-cinchona alkaloid used for strong binding of OsO_4*

References

1 BP 231901/1924 [*Chem. Abs.*, 1925, **19**, 3571].
2 M. Fetizon and M. Golfier, *C. R. Acad. Sci. Ser. C*, 1968, **267**, 900.
3 G.H. Posner, *Agnew. Chem., Int. Ed. Engl.*, 1978, **17**, 487.
4 A. McKillop and K.W. Young, *Synthesis*, 1979, 401 and 481.
5 'Polymer Supported Reactions in Organic Synthesis', eds. P. Hodge and D.C. Sherrington, Wiley and Sons, Chichester, 1980.
6 'Preparative Chemistry Using Supported Reagents', ed. P. Laszlo, Academic Press, San Diego, 1987.
7 J.H. Clark and D.J. Macquarrie, *Org. Proc. Res. Dev.*, 1997, **1**, 149.
8 D.C. Sherrington, in 'Chemistry of Waste Minimisation', ed. J.H. Clark, Blackie Academic and Professional, London, 1995.
9 P.T. Tanev and T.J. Pinnavaia, *Science*, 1995, **267**, 865.
10 C.T. Kresge, M.E. Leonowicz, W.J. Roth, J.C. Vartuli and J.S. Beck, *Nature*, 1992, **359**, 710.
11 J.H. Clark, A.J. Hyde and D.K. Smith, *J. Chem Soc., Chem. Commun.*, 1986, 791.
12 J. Ichihara, T. Matsuo, T. Hanafusa and T. Ando, *J. Chem. Soc., Chem. Commun.*, 1986, 793.
13 G. Hutchings, P. Page and F. Hancock, *Chem. Br.*, July 1997, 46.
14 J.H. Clark, P.M. Price, K. Martin, D.J. Macquarrie and T.W. Bastock, *J. Chem. Res.*, 1997, 430; EP Appl., 1998.
15 J.H. Clark, A.P. Kybett and D.J. Macquarrie, 'Supported Reagents: Preparation, Analysis and Applications', VCH, New York, 1992.
16 J.H. Clark and C.W. Jones, *J. Chem. Soc., Chem. Commun.*, 1987, 1409.
17 J.H. Clark and D.J. Macquarrie, *Chem. Commun.*, 1998, 853.
18 C.A. Muller, M. Maciejewski, T. Mallat and A. Baiker, *J. Catal.*, 1999, **184**, 280.
19 C.G. Flowers, S. Lindley and J.E. Leffler, *Tetrahedron Lett.*, 1984, 499.
20 S.J. Tavener, J.H. Clark, G.W. Gray, P.A. Heath and D.J. Macquarrie, *Chem. Commun.*, 1997, 1147.
21 J.H. Clark, 'Catalysis of Organic Reactions by Supported Inorganic Reagents', VCH, New York, 1994.
22 'Solid Supports and Catalysts in Organic Synthesis', ed. K. Smith, Ellis Horwood, Chichester, 1992.
23 R.J.P. Corriu and D. Leclercq, *Angew. Chem., Int. Ed. Engl.*, 1996, **35**, 1420.
24 A. Corma, *Chem. Rev.*, 1997, **97**, 2373.
25 D. Brunel, N. Bellocq, P. Sutra, A. Cauvel, M. Lasperas, P. Moreau, F. Di Renzo, A. Galarneau and F. Fajula, *Coord. Chem. Rev.*, 1998, **178–180**, 1085.
26 D.J. Macquarrie, D.B. Jackson, J.E. Mdoe and J.H. Clark, *New J. Chem.*, 1999, **23**, 539.
27 S.L. Burkett, S.D. Sims and S. Mann, *Chem. Commun.*, 1996, 1367.
28 S. Mann and C.E. Fowler, *Chem. Commun.*, 1997, 1769.
29 D.J. Macquarrie, *Chem. Commun.*, 1996, 1961.
30 R. Corriu, *C. R. Acad. Sci. Paris, Serie II c*, 1998, 83.
31 M. Onaka, K. Sugita and Y. Izumi, *J. Org. Chem.*, 1989, **54**, 1116.
32 J.H. Clark and C.V. Duke, *J. Org. Chem.*, 1985, **50**, 1330.
33 A. Al-Jazzaa, J.H. Clark and M.S. Robertson, *Chem. Lett.*, 1982, 405.
34 Y-Y. Huang, B-Y. Zhao and Y-C. Xie, *Appl. Catal. A*, 1998, **173**, 27.
35 T. Ando, S.J. Brown, J.H. Clark, D.G. Cork, T. Hanafusa, J. Ichihara, J.M. Miller and M.S. Robertson, *J. Chem. Soc., Perkin Trans. 2*, 1986, 1133.
36 J.H. Clark, S.R. Cullen, S.J. Barlow and T.W. Bastock, *J. Chem. Soc., Perkin Trans 2*, 1994, 1117.
37 T.W. Bastock and J.H. Clark, in 'Speciality Chemicals', ed. B. Pearson, Elsevier, London, 1992.

38 I.C. Chisem, J. Rafelt, M.T. Shieh, J. Chisem, J.H. Clark, R. Jachuck, D.J. Macquarrie, C. Ramshaw and K. Scott, *Chem. Commun.*, 1998, 1949.

39 I. Chisem, K. Martin, T. Shieh, J. Chisem, J.H. Clark, R. Jachuck, D.J. Macquarrie, J. Rafelt, C. Ramshaw and K. Scott, *Org. Proc. Res. Dev.*, 1997, 1, 365.

40 D.J. Macquarrie, *Green Chemistry*, 1999, 195.

41 T. Tatsumi, K.A. Koyano and N. Igarashi, *Chem. Commun.*, 1998, 325.

42 J.H. Clark, A.P. Kybett, D.J. Macquarrie, S.J. Barlow and P. Landon, *J. Chem. Soc., Chem. Commun.*, 1989, 1353.

43 J.H. Clark and D.J. Macquarrie, *Chem. Soc. Rev.*, 1996, 303.

44 C.N. Rhodes and D.R. Brown, *J. Chem. Soc., Faraday Trans.*, 1992, 88, 2269.

45 J.H. Clark, S.J. Tavener and S.J. Barlow, *Chem. Commun.*, 1996, 2429.

46 H. Hattori, O. Takahashi, M. Takagi and K, Tanabe, *J. Catal.*, 1981, 68, 132.

47 K. Tanabe, H. Hattori and T. Yamaguchi, *Crit. Rev. Surf. Sci.*, 1990, 1, 1.

48 C. Reichardt, 'Solvents and Solvent Effects in Organic Chemistry', VCH, New York, 2nd Edn., 1988.

49 J.C. Yori, L.M. Krasnogor and A.A. Castro, *React. Kinet. Catal. Lett.*, 1986, 32, 27.

50 H.E.B. Lempers and R.A. Sheldon, *J. Catal.*, 1998, 175, 62.

51 Y. Godberg, 'Phase Transfer Catalysis Selected Problems and Applications', Gordon and Breach, Yverdon, Switzerland, 1992.

52 G.D. Yadav and C.K. Mistry, *J. Mol. Catal.*, 1995, 102, 67.

53 C. Ramshaw, *Green Chemistry*, 1999, G15.1.

54 J.H. Clark, A.P. Kybett, P. Landon, D.J. Macquarrie and K. Martin, *J. Chem. Soc., Chem. Commun.*, 1989, 1355.

55 I.C. Chisem, K. Martin, M.T. Shieh, J. Chisem, J.H. Clark, R. Jachuck, D.J. Macquarrie, J. Rafelt, C. Ramshaw and K. Scott, *Org. Proc. Res. Dev.*, 1997, 1, 365.

56 T.C Zempty, R.A. Gogins, Y. Mazur and L.M. Miller, *J. Org. Chem.*, 1981, 46, 4545.

57 G. Cooke, V. Sage and T. Richomme, *Synth. Commun.*, 1999, 29, 1767.

58 R.S. Varma, D. Kumar and R. Dahiya, *J. Chem. Res.*, 1998, 324

59 H. Yoshida, C. Murata and T. Hattori, *Chem. Commun.*, 1999, 1551.

60 R. Stavrescu, T. Kimura, M. Fumjiya, M. Vinatoru and T. Ando, *Synth. Commun.*, 1999, 1719.

61 M. Inace, Y. Itoi, S. Enomoto and Y. Watanabe, *Chem. Lett.*, 1982, 1375.

62 E. Santaniello, in 'Preparative Organic Chemistry Using Supported Reagents', ed. P. Hodge, Academic Press, New York, 1987, chapter 18.

63 A. Butterworth, J.H. Clark, P.H. Walton and S.J. Barlow, *Chem. Commun.*, 1996, 1859.

64 I.C. Chisem, J. Chisem and J.H. Clark, *New J. Chem.*, 1998, 81.

65 J. Chisem, I.C. Chisem, J.S. Rafelt, D.J. Macquarrie and J.H. Clark, *Chem. Commun.*, 1997, 2203.

66 Y. Kurusu, *React. Polym.*, 1995, 25, 63.

67 Y. Kurusu and D.C. Neckers, *J. Org. Chem.*, 1991, 56, 1981.

68 Y.V. Subba Rao, D.E. De Vos, T. Bein and P.A. Jacobs, *Chem. Commun.*, 1997, 355.

69 J.E. Elings, R. Ait-Meddour, J.H. Clark and D.J. Macquarrie, *Chem. Commun.*, 1998, 2707.

70 T. Maschmeyer, F. Rey, G. Sauhar and J.M. Thomas, *Nature*, 1995, 378, 159.

71 M. Taramasso, G. Perego and B. Notari, *USP* 4 410 501/1982.

72 B. Notari, *Stud. Surf. Sci. Catal.*, 1988, 37, 413.

73 R.A. Sheldon, *Chemtech.*, 1994, 38.

74 J.H. Clark, *Green Chemistry*, 1999, 1, 1.

75 I.C. Chisem, J. Rafelt, M.T. Shieh, J. Chisem, J.H. Clark, R. Jachuck, D.J. Macquarrie, C. Ramshaw and K. Martin, *Chem. Commun.*, 1998, 1949.

76 H. Turk and W.T. Ford, *J. Org. Chem.*, 1991, 56, 1523.

77 P. Battioni, J.-F. Bartoli, D. Mnasuy, Y.S. Byun and T.G. Traylor, *J. Chem. Soc., Chem. Commun.*, 1991, 1051.
78 P.R. Cooke and J.R. Lindsay Smith, *J. Chem. Soc., Perkin Trans. 1*, 1994, 1914.
79 C. Gilmartin and J.R. Lindsay Smith, *J. Chem Soc., Perkin Trans. 2*, 1995, 243.
80 A.B. Sorokin and A. Tuel, *New J. Chem.*, 1999, **23**, 473.
81 R. Neumann and T. J. Wang, *Chem. Commun.*, 1997, 1915.
82 D.R. Patel, M.K. Dalal and R.N. Ram, *Stud. Surf. Sci.*, 1998, **113**, 293.
83 S. Suresh, S. Skaria and S. Ponrathnam, *Stud. Surf. Sci.*, 1998, **113**, 915.
84 R.I. Kuresthy, N.H. Khan, S.H.R. Abdi and P. Iyer, *React. Funct. Polym.*, 1997, **34**, 153.
85 J.R. Rafelt and J.H. Clark, *Catal. Today*, 2000, in the press.
86 A. Chakrabarta and M.M. Sharma, *React. Polym.*, 1993, **20**, 1.
87 G.A. Olah, in 'Acidity and Basicity of Solids. Theory, Assesment and Utility', eds. J. Fraissard and L. Petrakis, Nato ASI Ser., Kluwer Academic, Dordrecht, 1994, vol. 444, pp. 305–334.
88 P.B. Venuto, *Microporous Mater.*, 1994, **2**, 297.
89 A. Corma, *Chem. Rev.*, 1995, **95**, 559.
90 I.V. Kozhievnikov, *Catal. Rev. Sci. Eng.*, 1995, **37**, 311.
91 X. Song and A. Sayari, *Catal. Rev. Sci. Eng.*, 1996, **38**, 329.
92 M.A. Harmer, Q. Sun, M.J. Michalczyk and Z. Yang, *Chem. Commun.*, 1997, 1803.
93 P. Laszlo and A. Mathy, *Helv. Chim. Acta*, 1987, **70**, 577.
94 J.H. Clark, S.R. Cullen, S.J. Barlow and T.W. Bastock, *J. Chem. Soc., Perkin Trans. 2*, 1994, 1117.
95 D.R. Brown, H.G.M. Edwards, D.W. Farwell and J. Massam, *J. Chem. Soc., Faraday Trans.*, 1996, **92**, 1027.
96 J.H. Clark, S.R. Cullen, S.J. Barlow and T.W. Bastock, *J. Chem. Soc., Perkin Trans 2*, 1994, 411.
97 G.D. Yadav and T.S. Thorat, *Tetrahedron Lett.*, 1996, 5405.
98 S.R. Cullen, *D. Phil. Thesis*, University of York, 1993.
99 *Envirocats*, Contract Catalysts Ltd., Merseyside, England.
100 P.D. Clark, A. Kirk and J.G.K. Yee, *J. Org. Chem.*, 1995, **60**, 1936.
101 H.P. van Shaik, R.J. Vijn and F. Bickelhaupt, *Angew. Chem., Int. Ed. Engl.*, 1994, **33**, 1611.
102 J.H. Clark, J.C. Ross, D.J. Macquarrie, S.J. Barlow and T.W. Bastock, *Chem. Commun.*, 1997, 1203.
103 J.C. Ross, *D. Phil. Thesis*, University of York, 1999.
104 B.P. Bandgar, S.R. Jagtab, B.B. Aghade and P.P. Wadgaonkar, *Synth. Commun.*, 1995, **25**, 2211.
105 B.P. Bandgar, S.R. Jagtab, S.B. Ghodeshwar and P.P. Wadgaonkar, *Synth. Commun.*, 1995, **25**, 2993.
106 B.P. Bandgar, M.B. Zirange and P.P. Wadgaonkar, *Synlett*, 1996, 149.
107 S.P. Kasture, B.P. Bhandgar, A. Sarkar and P.P. Wadgaonkar, *Synth. Commun.*, 1996, **26**, 1579.
108 B.P. Bhandgar, C.T. Hajare and P.P. Wadgaonkar, *J. Chem. Res.*, 1996, 90.
109 T. Beregszaszi and A. Molnar, *Synth. Commun.*, 1997, 3705.
110 R.S. Drago and E.E. Getty, *J. Am. Chem. Soc.*, 1988, **110**, 331.
111 R.S. Drago and E.E. Getty, *Inorg. Chem.*, 1990, **29**, 1186.
112 R.S. Drago and E.E. Getty, *USP* 4 719 190/1988.
113 R.S. Drago and E.E. Getty, *USP* 4 929 80/1990.
114 J.H. Clark, K. Martin, A.J. Teasdale and S.J. Barlow, *J. Chem. Soc., Chem. Commun.*, 1995, 2037.
115 J.H. Clark, P.M. Price, K. Martin, A.J. Teasdale, D.J. Macquarrie and T.W. Bastock, *J. Chem. Res.*, 1997, 430.
116 J.H. Clark, P.M. Price, K. Martin, A.J. Teasdale, D.J. Macquarrie and T.W. Bastock, *EP Appl.*, 98302068.6/1998.

117 R. Ryo, S. Jun, J. Man Kim and M. Jeong Kim, *Chem. Commun.*, 1997, 2225.
118 R. Mokaya and W. Jones, *Chem. Commun.*, 1997, 2185.
119 K. Wilson and J.H. Clark, *Chem. Commun.*, 1998, 2135.
120 K. Wilson and J.H. Clark, *WP Appl.*, PCT/EP99/06529/1999.
121 B.M. Khadilkar and S.D. Borkar, *Tetrahedron Lett.*, 1997, 1641.
122 K. Tanabe, *Crit. Rev. Surf. Chem.*, 1990, 1, 1.
123 M.A. Harmer, Q. Sun, M.J. Michalczyk and Z. Yang, *Chem. Commun.*, 1997, 1803.
124 Q. Sun, M.A. Harmer and W.E. Farneth, *Chem. Commun.*, 1996, 1201.
125 M.A. Harmer, W.E. Farneth and Q. Sun, *Adv. Mater.*, 1998, 10, 1255.
126 W.M. Van Rhijn, D.E. De Vos, B.F. Sels, W.D. Bossaert and P.A. Jacobs, *Chem. Commun.*, 1998, 317.
127 K. Wilson and J.H. Clark, unpublished results.
128 J.M. Riego, Z. Sedin, J.M. Zaldivar, N.C. Marziano and C. Tortato, *Tetrahedron Lett.*, 1996, 37, 513.
129 Y. Izumi, K. Urabe and M. Onaka, *Microporous Mesoporous Mater.*, 1998, 21, 227.
130 Y. Izume, M. Ono, M. Ogawa and K. Urabe, *Chem. Lett.*, 1993, 825.
131 Y. Izume, M. Ono, M. Kitigawa, M. Yoshida and K. Urabe, *Microporous Mater.*, 1995, 225.
132 G. Suzukamo, M. Fukao and M. Minobe, *Chem. Lett.*, 1987, 58.
133 M.W. Branco, R.Z. Cao, L.Z. Liu and G. Ege, *J. Chem. Res. (S)*, 1999, 274.
134 J.H. Clark, *Chem. Rev.*, 1980, 80, 429.
135 M. Lasperas, N. Bellocq, D. Brunel and P. Moreau, *Tetrahedron Asymmetry*, 1998, 3053.
136 D.J. Macquarrie, J.H. Clark, A. Lambert, A. Priest and J.E.G. Mdoe, *React. Funct. Polym.*, 1997, 35, 153.
137 E. Angeletti, C. Canepa, G. Martinetti and P. Ventuerello, *J. Chem. Soc., Perkin Trans. 1*, 1989, 105.
138 E. Angeletti, C. Canepa, G. Martinetti and P. Ventuerello, *Tetrahedron Lett.*, 1988, 2261.
139 D.J. Macquarrie and D.B. Jackson, *Chem. Commun.*, 1997, 1781.
140 M. Lasperas, T. Lloret, L. Chaves, L. Rodriguez, A. Cauvel and D. Brunel, *Stud. Surf. Sci. Catal.*, 1997, 108, 75.
141 A. Cauvel, G. Renard and D. Brunel, *J. Org. Chem.*, 1997, 62, 749.
142 D.J. Macquarrie, *Tetrahedron Lett.*, 1998, 39, 4125.
143 I. Rodriguez, S. Iborra, A. Corma, F. Rey and J.L. Jorda, *Chem. Commun.*, 1999, 593.
144 P. Tundo, *J. Chem. Soc., Chem. Commun.*, 1977, 641.
145 P. Tundo, P. Venturello and F. Angeletti, *J. Am. Chem. Soc.*, 1982, 104, 6551.
146 Y.B. Subba Roa and B.M. Choudary, *Synth. Commun.*, 1992, 22, 2711.
147 A. Akulah and D.C. Sherrington, *Chem. Rev.*, 1981, 81, 557.
148 O. Orrad and Y. Sasson, *J. Org. Chem.*, 1990, 55, 2952.
149 P. Tundo and P. Venturello, *J. Am. Chem. Soc.*, 1981, 103, 856.
150 P. Tundo and P. Venturello, *J. Am. Chem. Soc.*, 1979, 6606.
151 J.H. Clark, S.J. Tavener and S.J. Barlow, *Chem. Commun.*, 1996, 2429.
152 J.H. Clark, S.J. Tavener and S.J. Barlow, *J. Mater. Chem.*, 1995, 5, 827.
153 S.J. Tavener, *D. Phil. Thesis*, University of York, 1997.
154 K. Ravi Kumar, B.M. Choudary, Z. Jamil and G. Thyagarajan, *J. Chem. Soc., Chem. Commun.*, 1986, 130.
155 M.-Z. Cai, C.-S. Song and X. Huang, *Synthesis*, 1997, 521.
156 M.-Z. Cai, C.-S. Song and X. Huang, *Synth. Commun.*, 1997, 27, 361.
157 M.-Z. Cai, C.-S. Song and X. Huang, *J. Chem. Soc., Perkin Trans. 1*, 1997, 2273.
158 M.-Z. Cai, C.-S. Song and X. Huang, *Synth. Commun.*, 1998, 693.
159 B.F.G. Johnson, S.A. Raynor, D.S. Sheppard, T. Mashmeyer, J.M. Thomas, G. Sankar, S. Bromley, R. Oldroyd, L. Gladden and M.D. Mantle, *Chem. Commun.*, 1999, 1167.

160 T.J. Pinnavaia, R. Raythatha, J.G-s. Lee, L.J. Halloran and J.F. Hoffman, *J. Am. Chem. Soc.*, 1979, **101**, 6891.

161 T.J. Pinnavaia, *Science*, 1983, **220**, 365.

162 R. Augustine, S. Tanielyan, S. Anderso and H. Yang, *Chem. Commun.*, 1999, 1257.

163 J.M. Thomas and W.J. Thomas, 'Principles and Practice of Heterogeneous Catalysis', VCH, Weinheim, 1997.

164 M.D. Ward, T.V. Harris and J. Schwartz, *J. Chem. Soc., Chem. Commun.*, 1980, 357.

165 M. McCann, E.M. Coda and K. Maddock, *J. Chem. Soc., Dalton Trans.*, 1994, 1489.

166 E. Lindner, A. Jager, M. Kemmler, F. Auer, P. Wegner, H.A. Mayer and E. Plies, *Inorg. Chem.*, 1997, **36**, 862.

167 C.E. Song, C.R. Oh, S.W. Lee, S-g. Lee, L. Canali and D.C. Sherrington, *Chem. Commun.*, 1998, 2435.

168 C.E. Song, J.W. Yang and H.J. Ha, *Tetrahedron Asymmetry*, 1997, **8**, 841.

169 P. Gros, P. Le Perchec and J.-P. Senet, *J. Chem. Res.*, 1995, 196.

Subject Index